The Practical Guide to Simple Structures

Brian R Sheppard

B.Sc., C.Eng., M.I.C.E., M.I.Struct.E., M.I.A.S.

Coterie

First published in Great Britain 1983 by
Coterie
Old Road, Maisemore, Gloucester GL2 8HS

British Library Cataloguing in Publication Data
Sheppard, Brian R.
The Practical Guide to Simple Structures
1. Structural Engineering
I. Title
624 TA633
ISBN 0-950861-00-6

ISBN 0 9508610 0 6

Phototypesetting by Action Typesetting, Gloucester.
Printed by Orchard and Ind Ltd., Gloucester.

acknowledgments

I have received and am therefore pleased to acknowledge the encouragement and valuable advice and comment received from Derek T. Preece, B.Sc., R.I.B.A., F.F.A.S., F.G.S., during the final preparation stages of this book.

Extracts from British Standards are reproduced by permission of the British Standards Institution, 2, Park St., London W1A 2BS from whom complete copies can be obtained.

Contents

Chapter Four

Chapter Five

Appendix

ERRATA

page 15 " eaves to ridge of 3.000 m "
" 82 x 3.000/2 = 123 kg/m "
page 48 " D/T for this beam is 11.2 "
page 59 " 733 mm² "
page 72 " 6.33 mm "
page 86 " line 56 - 2/3 the thickness "

Introduction

Even you can cope with structures

If you are someone who has difficulty with structures, I hope you will accept the challenge and progress with me through this book. If you can cope at the moment, I hope you will also proceed and benefit from perhaps this more fundamental approach to the subject. For those of you who have gone through the "mill" I think you will find this a help to simple structures at a professional and commercial level and maybe a consolidation of your basic knowledge.

There is no mystery to structures, but anything one does not understand can often be mysterious until the light dawns. If you are in the dark it is possibly because academic courses cannot by their nature and their direction towards examinations spend as much attention to the "why" and the "how" of basic principles as some students need. If you do not grasp square one therefore, square two and those that follow proceed in varying degrees of paralysis and fog.

Those of you who are not paralysed will, I hope, become more fluent in the way we resist the forces of gravity in the building process and perhaps add a few more weapons to your armoury of skills in this direction.

Having progressed through understanding I look at the professional and commercial side of calculations. The way we put our knowledge into practice through calculations is seldom taught and is an important element in itself for proper and efficient evaluation of new and existing structures. Be in no doubt, that even if you are only involved in single beams every now and then, doing things the proper way is very important and will assist accuracy, your own understanding and that of the reader of your work.

To help the calculating process, I have included relevant tables from British Standards and Codes of Practice to cover most of the work needed for the student or practitioner in this one publication.

This is not an academic treatise into the precise way structures behave and intereact. There are many other excellent works dealing with those matters in varying depths.

The title of this book has been carefully chosen to reflect its contents. It is primarily a practical and a practice book. There is no short cut to a sound understanding of structures but with the objects of this book in mind, I have tried to condense the theory to the minimum sufficient to cope with those matters which comprise the majority of structural problems that arise in an architects or builders office.

It would have been easier but quite wrong to produce an even more potted version of building structures which would probably have enabled most people to produce solutions in a copying way and parrot fashion. This would have been dangerous because it is essential to be aware, even roughly, of what is happening in your structure, so that common sense, that most excellent of supervisors, can play its proper part.

I wish you well then as we proceed.

Chapter One

What is structure

We must realise to start with that STRUCTURE is the way in which natural materials resist the natural forces of gravity and wind. We are particularly interested here in how this is done in buildings. In this context we may consider buildings as man-made assemblies of natural materials made to provide shelter, warmth or some other amenity.

It is important then to see how such work is linked with nature.

An important structure in nature itself which we can all understand is the tree. To me the tree is a growing display of perfect structure. In structural terms the branches are arms or cantilevers transmitting the wind and gravity forces from the leaves and twigs to the main trunk. The trunk itself is a vertical arm or cantilever getting thicker towards the ground as the loads increase. Once in the ground the trunk subdivides back again to smaller roots which, by their hold in the soil, transfer all these loads into the ground.

There are four important points to see in the tree, the significance of which we shall see later. Firstly, the branches act as the small scale elements transferring loads horizontally to the trunk. Secondly, the trunk is the main vertical element and weight carrying unit. Thirdly, the root system provides the foundation, allowing the loads from the superstructure to be satisfactorily transmitted to the ground. Lastly and the main lesson to be learnt is that all three have to be present **all the time**. This last point may be obvious but is essential to keep in mind. It is all too easy for the inexperienced to forget it when we get in amongst a building.

From nature we can progress to man-made structures.

One of the earliest man-made structures is the "post and beam" bridge. Some can still be seen in the lake district and Devon and are illustrated in Figure 1.

These bridges were made out of solid stones, quarried and laid in position as blocks. It is important to see that the stone is being used horizontally and is supported at both ends as a beam. Nowhere does this happen in nature and is the significant difference between materials in their natural state and the use of them by man in what we know as STRUCTURE.

Figure 1

The structures with which we shall be dealing in this book do not vary in principle from the "post and beam". They do vary of course in detail, in their complexity and in that they are man-made.

terms and definitions

At this stage we need to arm ourselves with some terms and definitions. The first few are easy and are little more than clarification to the student of certain building construction terms which have structural significance. The rest are more difficult and involve the first of the structural concepts that you have to grasp before proceeding further.

the easy ones...

A SUPERSTRUCTURE is that part of a building or construction that is above ground.

A FOUNDATION is that part of a building which, like the roots of the tree, transfer the weight and loads of the superstructure to the ground.

A LOAD-BEARING WALL is a wall which is carrying other loads

(e.g. from floors and roofs) and is therefore supporting more than its own weight.

A COLUMN is an item of vertical support like the wall. The difference being that its support is at a point and not along a line.

A BEAM is a general structural term for a horizontal element which carries load across two or more points of support. You will see below that the general word BEAM is given different particular names when used in certain places.

RAFTERS are beams in the sloping part of a roof which carry the roof deck itself at close and regular spacing.

A PURLIN is a beam in the sloping part of a roof providing intermediate support to rafters by being underneath them and spanning in the other direction.

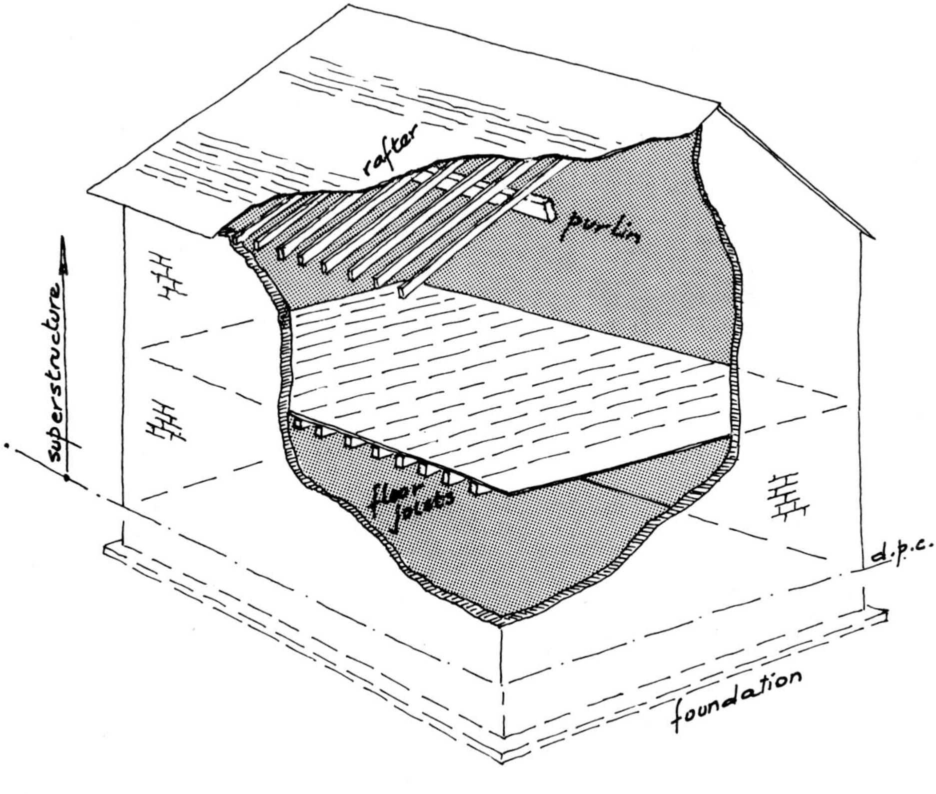

Figure 2

A JOIST is the more general description of a beam in building construction. They can be found singley or laid at close and regular spacing as in a floor or ceiling to support the floor boarding deck or ceiling skin.

A BEARING is that part of a wall which provides support for a beam.

For clarity these are now all illustrated in Figure 2.

and some more difficult...

Sir Isaac Newton discovered that there is a natural attraction between two bodies. This attraction is called FORCE. This explained GRAVITY with the two bodies concerned being the earth and the apple that fell on his head. This force is greater the closer together the two bodies become. As explained earlier, all structure is the behaviour of materials under the action directly and indirectly of this gravitational force.

The UNIT of FORCE is the NEWTON (N)

1 kilonewton (kN) = 1 000 Newtons

1 kN = 1 000 N

WEIGHT is the force which gravity causes to act on bodies. It is proportional to the distance the body is away from the centre of the earth and disappears when that same body is out in space virtually free from the gravitational force of attraction. That body then becomes **Weightless.**

MASS is the property of a body which does not change however far away from the earth. The unit of mass is the GRAM (g) and KILOGRAM (kg)

1 kg = 1 000 g

Clearly WEIGHT and MASS are directly related but as we have seen they vary with the distance above the earth. The small variations experienced on the ground however, are negligible and it is generally accepted that 1 KILOGRAM MASS produces a downward **weight** under gravity of just over 9.8 NEWTONS force. I shall now simplify this for the purposes of this book and the size of structures we shall be dealing with to the following,

1 kg = 10 N

Do not forget however, the difference between MASS and WEIGHT force because, even though we may write the above equation and think of it for convenience, a mass can never **equal** a weight but **produces** it. The kilogram is mass and the newton is the unit force of weight. We may finish this definition with the simple development from the above that,

$$\mathbf{100\ kg = 1\ 000\ N = 1\ kN}$$

For equilibrium a force has to act on something and therefore be resisted by an equal and opposite force. The effect it has on the object is called STRESS and this stress will vary with the size or area over which the force is acting. STRESS is a FORCE PER UNIT AREA. In other words it is the amount of force being carried by each square of unit area over which it is acting. Where these units of area are SQUARE MILLIMETRES, the UNITS of STRESS become NEWTONS PER SQUARE MILLIMETRE or N/mm^2.

This is illustrated in Figure 3.

In the first example the 1000 kg. MASS which produces a WEIGHT force of 10 000 N is acting uniformly on a block of 200 mm × 100 mm or an AREA of 20 000 mm^2. It is, therefore, producing,

$$\text{a STRESS of } \frac{10{,}000\ N}{20{,}000\ mm^2} = 0.5\ N/mm^2.$$

In the second example the same MASS and WEIGHT are acting on a smaller block of 100 mm × 50 mm and the STRESS is therefore,

$$\frac{10{,}000\ N}{5{,}000\ mm^2} = 2.0\ N/mm^2.$$

The UNIT of STRESS is therefore FORCE/UNIT AREA. In the illustration this is N/mm^2. With a little careful thought and a pencil you will see that,

$$\mathbf{1\ N/mm^2 = 1\ 000\ 000\ N/m^2 = 1\ 000\ kN/m^2.}$$

We can now look at two more units which follow on from the above that you will have heard of but which we will not be using much in this book of simple structures.

STRAIN is the **effect** of stress on materials. Now this will vary with the **type** of material and not its **size** and is the amount by which it changes in shape under stress. This is expressed as a ratio on the original length and in building materials is very, very small. To illustrate the point you can imagine that a column of suet pudding will change its shape quite considerably under a very small stress but a column of wood of the same

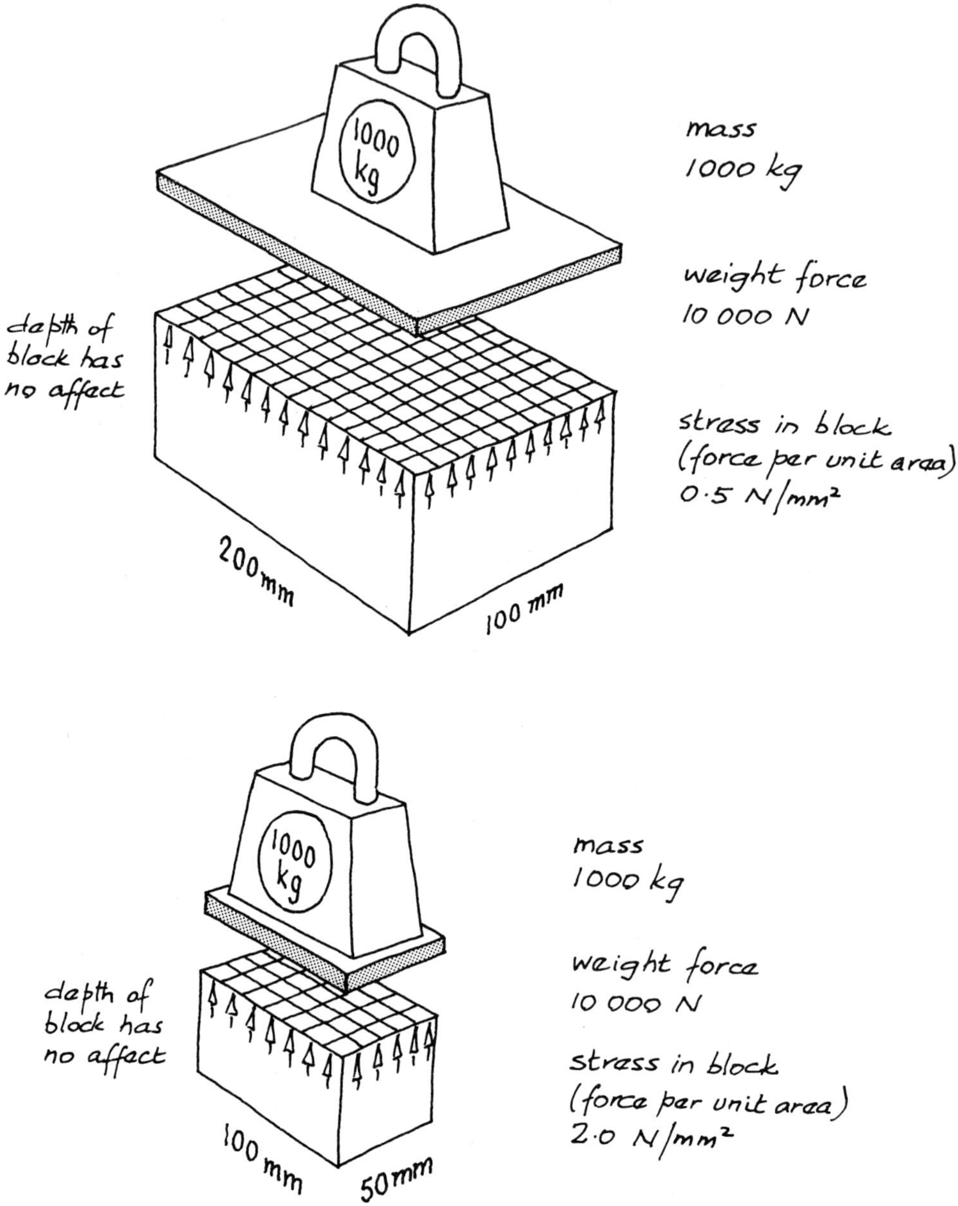
1000 kg
mass
1000 kg
weight force
10 000 N
depth of
block has
no affect
stress in block
(force per unit area)
0·5 N/mm²
200 mm
100 mm
1000 kg
mass
1000 kg
weight force
10 000 N
depth of
block has
no affect
stress in block
(force per unit area)
2·0 N/mm²
100 mm
50 mm

Figure 3

size will not really change its shape at all under the same stress. This is one reason why suet pudding is not a building material.

DEFLECTION is an example of strain or the effect of stress and is something we shall meet. DEFLECTION is the amount by which a beam will sag below the horizontal when spanning between two supports. We shall see that the amount of deflection will vary in different materials under the same loads by virtue of the STRAIN property of that material.

Chapter Two

Loads
and the stresses they induce

from bedroom to bedrock

We now know that gravity acts on all bodies, and that covers the building itself and everything in it. It is the job of the STRUCTURE to keep things where they are meant to be and not end up in a pile of rubble on the ground. Everything in a building then has an affect on the forces through the building and which the foundations eventually transmit to the ground.

The way these forces pass through and are resisted by the building elements is therefore worth looking at. For instance, floor joists in a floor, span in one direction and you would be right in thinking that the loads carried by that floor are in turn carried by the walls which provide the bearing to these floor joists. All the loads including the weight of the floor itself are carried by these end walls and nothing at all by the side walls parallel with the joists.

To illustrate this we have our simple building again in Figure 4.

We see that Dad has taken up brewing his own beer and much to Mother's annoyance has a line of 20 litre bins of the brew in the bedroom.These are 500 mm apart and, happily for this example, the line is across the joists at exactly mid-span. Each bin is of 25 kg. weight and therefore, for every metre along the line of bins there is a weight of 50 kg. i.e. 50 kg/m.

From what we saw in the last chapter the force which these bins are applying to the floor is, therefore, 500 N/m. As the bins are at mid-span

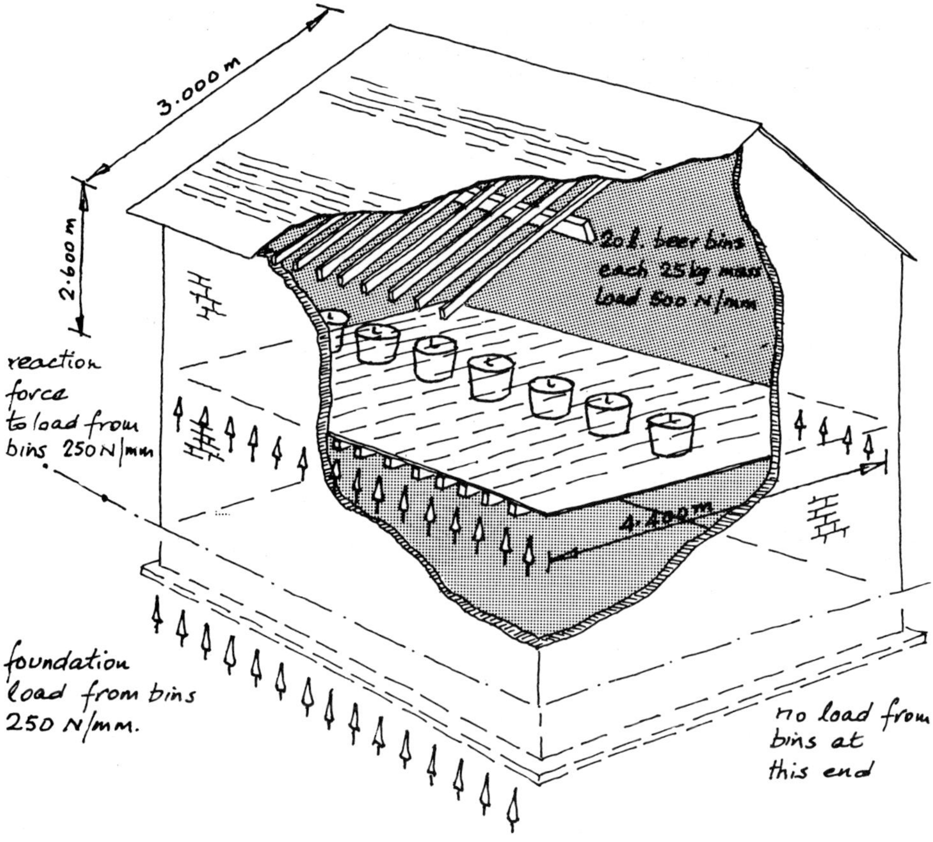

Figure 4

of the joists the resisting reaction to this force is equal at each end bearing. The bearing reaction force at each end is, therefore, half of the force being applied at mid-span, i.e. 250 N force in every metre along these end load-bearing walls (250 N/m). This resisting force to that being applied by the bins is itself supported all the way up from the wall foundations. The extra load into the end wall foundations arising from the bins of beer is, therefore, 250 N/m.

You will note that the foundation to the side walls experience no load from the floor or the bins in the bedroom and are quite oblivious of Dad's activities.

This loading from Dad's brewing bins is called LIVE LOAD. This is load from anything which is not part of the building carcass or fabric, the weight of which is called DEAD LOAD. In practice we do not need to know about Dad's activities as we have recommended LIVE LOADS per square metre for rooms of different uses. These are set down in a British Standard Code of Practice for Loading and most that you will meet are summarised in the Appendix. In this particular case it is 153 kg/m^2 over all the floor area.

dead load and its calculation

The dead load of the floor itself and the roof behaves in exactly the same way as the live load with the load going only in the direction of the joists or rafters respectively.

Dead Load, or the weight of the building fabric and materials is as much a contributor to the total loads through a building as are live loads. To calculate the total weight or dead load of a building is a daunting prospect. However, we are never involved with doing precisely that. What we are involved with is calculating the dead load (and live load) at a particular point in a structure or along a particular line where our structural problem or exercise exists. This is much more manageable as the calculation is limited to those dead and live loads above that position and which are passing through it.

Nevertheless there is a proper and business-like way of setting down even these calculations which will both save a lot of time and be accurate.

To calculate the total weight of a floor or a wall with all its constituent parts also seems an ungainly matter. However, the method to be used is a unitary method which you will see makes it all much easier.

In this we take a square metre of floor, or wall, and calculate the dead weight of that, adding up the various parts of the construction (per square metre) and achieving the unit weight (per square metre) of the floor or wall. Most sheet materials are expressed from catalogues in kilograms per square metre and other items such as timber, brick etc., are scheduled in the appendix in this form for easy use. Items not included can easily be worked out on this basis.

Let us assume the floor carrying the bins of beer is constructed of 18 mm floor boarding, 250 × 50 timber joists at 400 mm centres and has a ceiling below of 13 mm plasterboard and skim and that the span of the floor is 4.400m. (The weights used can be found in the Appendix)

The unit weight calculation of that floor would be set out as follows:

18 mm floor boarding	= 10 kg/m²
250 × 50 joists at 400 mm crs.	= 16
13 mm pbd and skim	= 22
	= 48 kg/m²

Note two points from this calculation. The first is that we always work from top to bottom through a unit and through a building in calculating loads. The second is a careful attention to and indication of the units in use. In a calculation like this it is a good convention to mark the units on the first and last of such a series of similar items.

The next step is to see the amount of the dead load of the floor which is transferred to and carried by the end load-bearing wall. As the weight of the floor is constant throughout its length half the span will be carried by one end and half by the other. The weight of this floor therefore at one end wall **per metre length of the wall** is,

$$48 \text{ kg/m}^2 \times 2{\cdot}200 \text{ m} = 105{\cdot}6 \text{ kg/m.}$$

Notice how when "kg/m²" are multiplied by "m" the resultant unit is "kg/m" — and that it has been marked accordingly.

The unit weight of an external wall would be similarly calculated. For a cavity wall of brick and 100 mm thick insulating blockwork with 13 mm of plaster internally the unit weight would be as follows:

100 mm brickwork	= 225 kg/m²
100 mm blockwork	= 60
13 mm plaster	= 22
	307 kg/m²

For a wall, the extension of this figure into kg/m at a particular position, is based on the height of wall above the point. Let us say here that it is 2·600 m between the bedroom floor and the eaves. Then the weight of wall being carried at the level of the bedroom floor is,

$$307 \text{ kg/m}^2 \times 2{\cdot}600 \text{ m} = 798{\cdot}2 \text{ kg/m.}$$

Pitched roofs can be a little more complicated in their action and are worthy of a separate explanation.

The weight of a simple pitched roof as in Figure 5 works as follows.

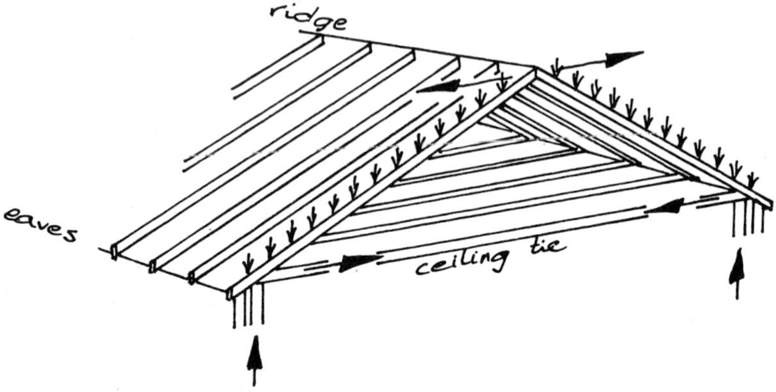

Figure 5

Whilst the roof rafter is spanning from eaves to ridge and is therefore gaining support at each end of its length, the ridge support can only be a horizontal one of thrust from the other slope of the roof. This horizontal thrust is balanced at the eaves usually by ceiling rafters acting as ties. In the absence of such balancing tie, the wall itself has to carry the sideways thrust which it can do to only a limited extent and the lack of a tie itself can cause problems which is a matter of more general building design. As there is nothing at the ridge to carry vertical loads all the load of one side of the pitched roof has to be carried at the eaves by the only item of vertical support which is the external wall.

Introduction of a purlin half way up the slope does not alter the logic but does reduce the load at the eaves. Figure 6 shows the situation and as

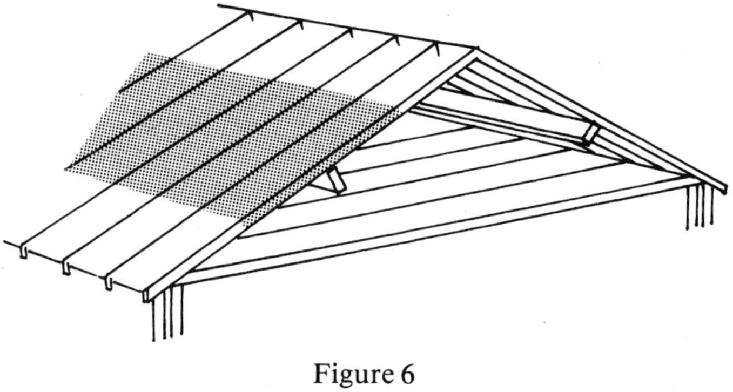

Figure 6

you may imagine the weight of the shaded area carried by the purlin is carried by those walls which support the purlin and not the eaves. Both the remaining part of the roof down to the eaves **and** that up to the ridge are carried by the wall at the eaves. In summary then, a purlin mid-way up the slope of the roof will carry half of the weight of that slope, the other half will be carried at the eaves. The flow of forces round a framed roof are in fact complex with many variables, but for the purposes of this book and the scale of calculations expected the above simplification is quite satisfactory and valid.

With a roof construction in Figure 4 of plain clay tiles on 25 × 12 mm battens on lining on 125 × 50 roof rafters the calculation would be as follows:

plain clay tiles	= 70 kg/m²
battens and lining	= 4
125 × 50 roof rafters	= 8
	82 kg/m²

When transferring this to an end support we must remember to measure the length up the slope and not horizontally. Because of the purlin the eaves load will be half that of the whole slope which has a length from eaves to ridge of 6·000 m. Roof dead load at one eaves support will therefore be,

$$82 \times 3{\cdot}000 = 246 \text{ kg/m}$$

There will be live load on the roof as a snow allowance which you can see added in Figure 7.

Working downwards from the top as I referred to before we can easily pick up the ceiling weight as follows:

125 × 50 ceiling rafters	= 8 kg/m²
plasterboard and skin	= 22
	30 kg/m²
Weight from ceiling on to eaves = 30 × 2·200	= 66 kg/m.

Whilst I have chopped up these calculations to add in explanation, the same work in practice would look as in figure 7. This would apply in the situation of an opening in the end wall over which would be needed a joist at say just below first floor level.

By inspection of the building layout and construction you will always have to assess which parts are transmitting load through the point that you are concerned with and then proceed with dead and live loads as above. for instance if the roof in Figure 4 had been spanning the other

To calculate load in side wall of Figure 4 just below first floor level

load from roof:	plain clay tiles	= 70 kg/m²	
	battens & lining	= 4	
	125×50 roof rafters	= 8	
		82 kg/m²	
	live load (snow) BSCP3	= 75	
		157 kg/m²	
	load on to eaves	= 157 × 1·500	= 235 kg/m
load from ceiling:	125×50 rafters	= 8 kg/m²	
	pbd. & skim	= 22	
		30 kg/m²	
	load on to eaves	= 30 × 2·200	= 66
load from wall:	100 mm brickwork	= 225 kg/m²	
	100 mm blockwork	= 60	
	13 mm plaster	= 22	
		307 kg/m²	
	load at floor level	= 307 × 2·600	= 798
load from floor:	18 mm floor boarding	= 10 kg/m²	
	250×50 floor joists	= 16	
	13 mm pbd & skim	= 22	
		48 kg/m²	
	live load BSCP Chap3	= 153	
		201 kg/m²	
	load at wall	= 201 × 2·200	= 442
	Total load at wall just below floor level		= 1541 kg/m
			= 15.41 kN/m

Figure 7

way then no load from it would be transmitted through the wall into which our joist was being placed, and that part of the dead and live loads would not have been added into our calculation. By the same token you must watch when considering some internal walls which are supporting purlins that you have included, therefore, that part of the roof load applicable. This is why you must always start at the top of a building in assessment of load and see which parts are being carried through your particular point and which are not.

Now that we can work out the load arriving at a particular location, the next part of the calculation is to carry that load across the opening concerned. To do that we need to know in more detail how the beam or horizontal element behaves under load.

the beam or horizontal element

In chapter one we saw how the beam was the first step that man took towards using natural material as structure. He did this by trial and error of course, noticing that for the span involved that particular stone did not break and further that he was able to walk over it.

We must now see why that is and look at what forces and stresses are acting inside the beam to allow this to happen. We know they must be there because gravity would otherwise pull the load being carried straight down through the middle.

Let us first go back to some more of Sir Isaac Newton's deductive logic. This was that for equilibrium to exist every force must have an equal and opposite force resisting it. My example of the short column in the definition of stress is easy enough to understand in this respect. There is a little more going on in a beam which we will now examine.

Figure 8(a) shows a beam, let us say a timber joist, being subjected to two equal point load forces at mid-span. A development from Sir Isaac Newton's theory is that if a system is in equilibrium (in other words it is stationary) every part of that system is also in equilibrium. We will take an imaginary power saw then and cut through the beam exactly in the

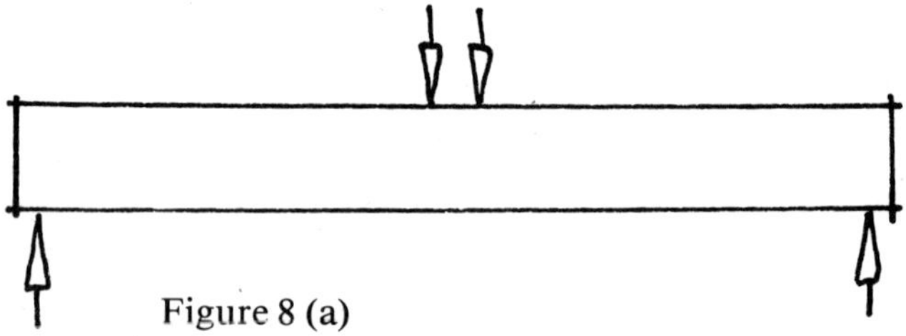

Figure 8 (a)

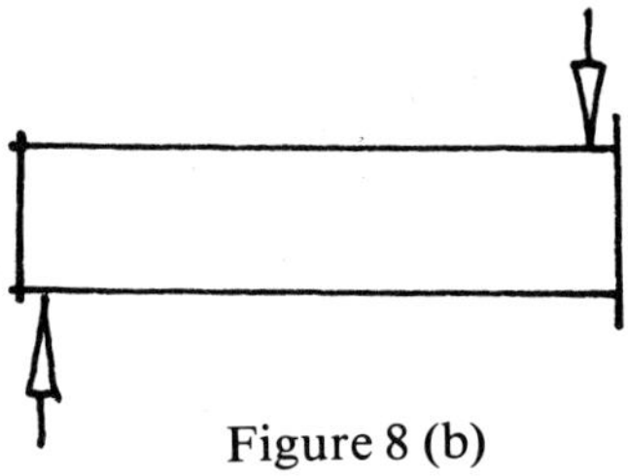

Figure 8 (b)

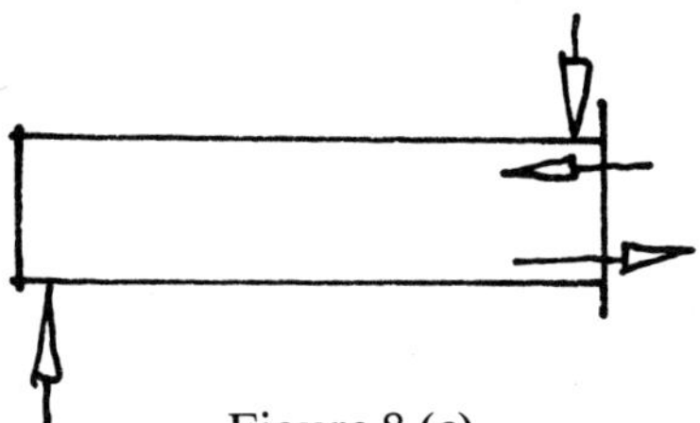

Figure 8 (c)

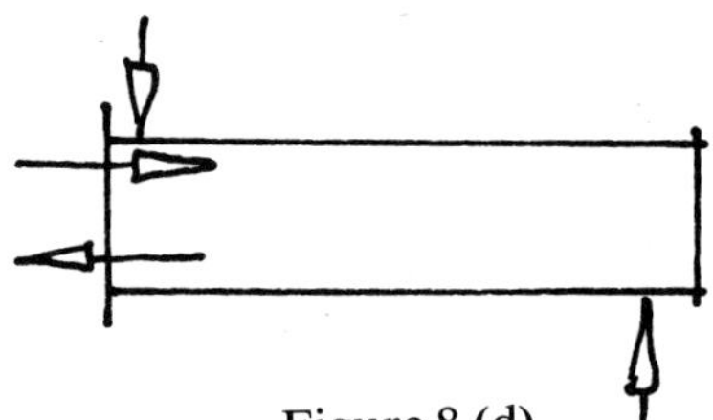

Figure 8 (d)

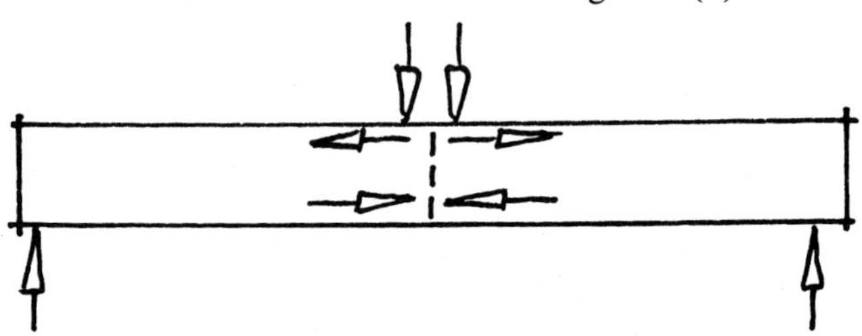

Figure 8 (e)

middle between the two loads. We then have the situation shown in figure 8(b). Here we see why the cut has to be imaginary as otherwise equilibrium would not be maintained and everything would fall down. This is because a real saw cut would have destroyed those connection forces within the beam across the cut line which were maintaining equilibrium and which we want to examine.

The two forces shown in figure 8(b) obviously have a turning effect on this bit of beam or in other words, if left to their own devices they would turn or rotate that bit of beam into collapse. This turning effect or tendency of the two forces to rotate is called a MOMENT. All equal, opposite and parallel forces have a turning or twisting action as we would use in unscrewing a lid off a jar and this is known in mathematics as a COUPLE. The amount of turning effect can be evaluated and it is the value of **either** of the equal forces multiplied by the distance between their lines of action. Therefore, if either the force or the distance between them is increased or decreased so will the turning effect or the MOMENT of the COUPLE.

The units of a moment or a BENDING MOMENT, as it is called in a beam, are therefore a bit unusual as they are a force multiplied by a distance. No creation of a new unit here though and the answer in our metric units is simple — the NEWTON MILLIMETRE (Nmm). Please take careful note that these units are being **multiplied** together and not **divided** as they were with stress (N/mm^2).

This moment then will have to be balanced for equilibrium by a turning effect of forces inside the beam in the other direction. This is called the MOMENT OF RESISTANCE and the forces can only be as shown in Figure 8(c) which are acting within the beam across the cut line. Further we can see that, if, for equilibrium, the value of each moment has to be the same, then because the internal forces are closer together than are the external ones their individual value must be greater in order to achieve a balance and maintain equilibrium.

We might just as easily have been considering the other half of the beam and if we had we would have ended up with the situation shown in Figure 8(d).

Now comes the clever bit because we will reconnect the two halves of our beam with the forces shown and we now have Figure 8(e).

We can now see the brilliant answer to all this because the forces in the top half of the beam are pushing outwards and are therefore resisting being squashed or COMPRESSION and the forces in the bottom half are pulling inwards and are resisting being stretched or TENSION.

This is the simply supported beam in all its glory with the two sets of forces at right angles to each other maintaining equilibrium and the peaceful life.

Whilst I positioned the imaginary cut in a position and circumstances to best suit the explanation, a cut somewhere else in the beam will still provide a logical explanation and will tell us a bit more about these internal forces.

Figure 9(a) shows the situation and there is something lacking to maintain equilibrium. The answer of course, must be another vertical force inside the beam and this is shown in Figure 9(b). This is a real force and is called SHEAR FORCE. This time by the addition of a **single** force the equilibrium is maintained.

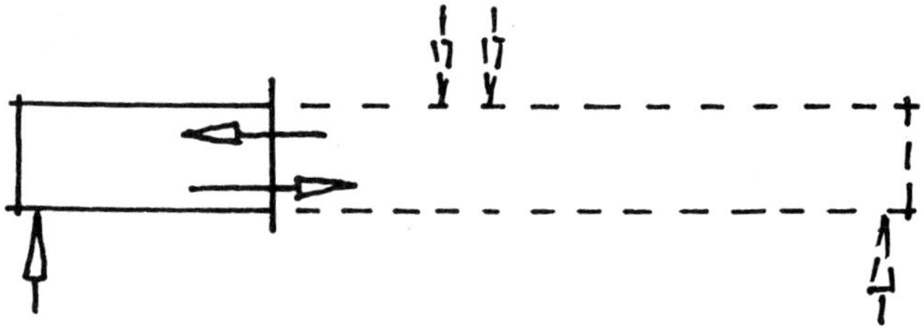

Figure 9 (a)

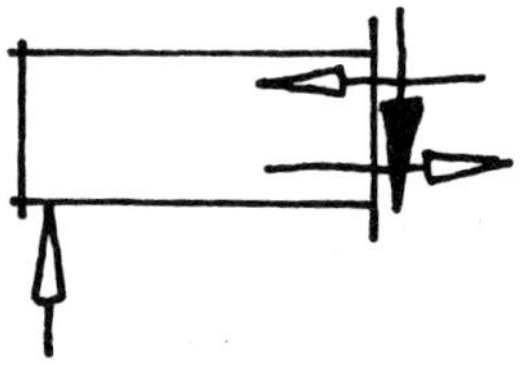

Figure 9 (b)

To summarise this most important single element of structure, the beam, we find that inside every beam under load there are two systems of forces acting. These are the two equal and opposite forces providing a MOMENT OF RESISTANCE and the individual vertical force or SHEAR FORCE.

bending moment

The strength of a beam is governed by its ability to resist the tension and compression forces inside it. As we have seen these create a turning effect or moment of resistance inside the beam which is equal to the external bending moment being applied. The principle of design is, therefore, to see what is the MAXIMUM BENDING MOMENT from the loads being applied in a particular case and from there to deal with the internal tension and compression created.

We will look then at how to calculate this bending moment and assess its maximum value.

a single point load

We saw in Figure 8 how the moment of compression and tension forces inside the beam had to balance the external loads; the greater the external load, the greater the compression, tension and shear force they create inside the beam for equilibrium to be maintained. In Figure 8(c) we deduced that it was the turning effect or moment that had to be balanced from outside to inside the beam, i.e. the force × the distance between them. This leads to a see-saw situation in that if the forces are closer together they need to be larger to have the same moment and vice-versa. The see-saw is a good analogy, because for an equal turning effect and therefore equilibrium a 9 stone man would need to be twice as far away from the pivot as an 18 stone man on the other end.

Look again at Figure 9. The value of the reaction support force carrying the end of the beam is obviously constant and hasn't changed during all this "cutting" and therefore, for equilibrium, neither has the balancing shear force. As the distance between them has reduced in Figure 9 from Figure 8 so must have their MOMENT. So also therefore has the balancing moment of compression and tension at that cut point. The important conclusion that has to be drawn from this is that in a beam with individual point loads and supports, the compression and tension force or moment of resistance produced by the bending moment **varies** along the length of the beam.

We shall now see how this can be evaluated and it is quite simple. To help us, let us give our example some meaning and call the mid-span point load W and the length of the beam L. This is now shown in Figure 10(a).

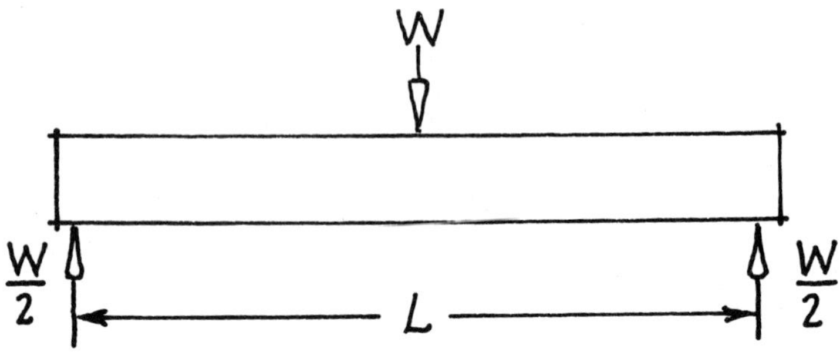

Figure 10 (a)

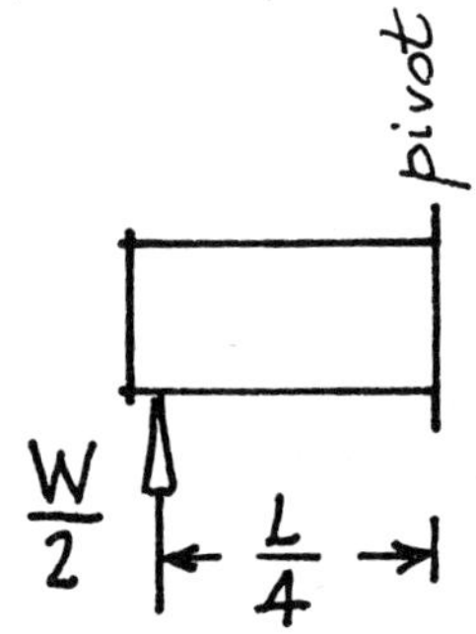

Figure 10 (b)

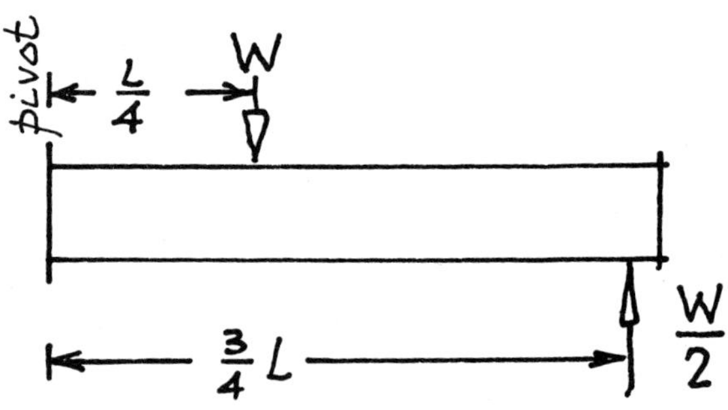

Figure 10 (c)

Let us now do what we did in Figure 9 and take our cut at the quarter point of the span as in Figure 10(b).

The turning effect or bending moment of the external forces at the point of the cut in Figure 10(b) is the reaction force at the end of the beam, W/2, multiplied by the distance to the cut line or to the pivot of

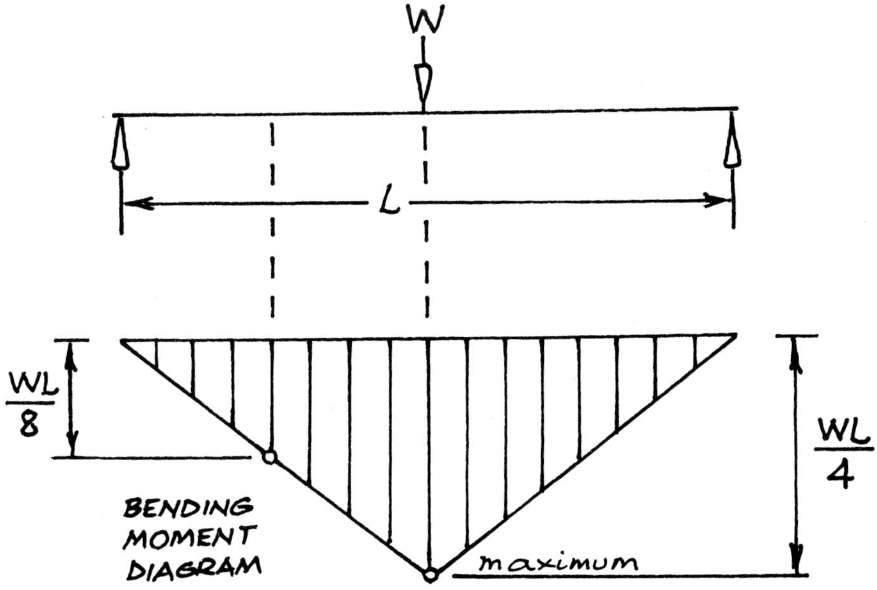

Figure 11

this particular see-saw which is L/4. This produces a BENDING MOMENT at the point of the cut of W/2 × L/4 or WL/8.

It should, of course, be the same if we had taken the same cut point and considered the other half of the beam as in Figure 10(c). Let us see. The turning effect of the external forces here are acting in opposite directions and so we shall have to subtract one from the other to get a resultant turning effect. W/2 × ¾L is the turning effect from the other support reaction, i.e. 3/8 WL, and that of the imposed point load is W × L/4 or 1/4 WL. The nett effect or bending moment at the cut is therefore,

$$3/8\ WL - 1/4\ WL = WL/8$$

which is what we had at the same point considering the other end of the beam. The moral of this story is that when you want to find out the bending moment at a point in a beam always calculate the simplest end.

To complete the picture for this beam we will now see what the bending moment is at two other points, at mid-span and at the end.

At mid-span (for one half of the beam),

$$\text{bending moment} = W/2 \times L/2 = WL/4$$

At one end,

$$\text{bending moment} = (W/2 \times L) - (W \times L/2) = 0.$$

Notice here that when we take bending moments at a point through which an external force is acting it has no turning effect about that point and therefore does not appear in the calculation. In other words its turning effect about the point is zero and therefore there is no moment arising from its action at that point.

We can plot the figures for bending moments we have just worked out diagrammatically as shown in Figure 11 and connect them up to provide a graph of the values along the beam.

This is called a BENDING MOMENT DIAGRAM and for a single point load it is triangular in shape. It enables the value of bending moment at any point to be read off. These have more relevance on more complex structures and in any event one is usually only interested, for design, in the peaks and the troughs. The maximum bending moment in this case is at the position of the point load and is WL/4.

We must now look at what happens when the point load moves away from the mid-span position and carry out the same procedure.

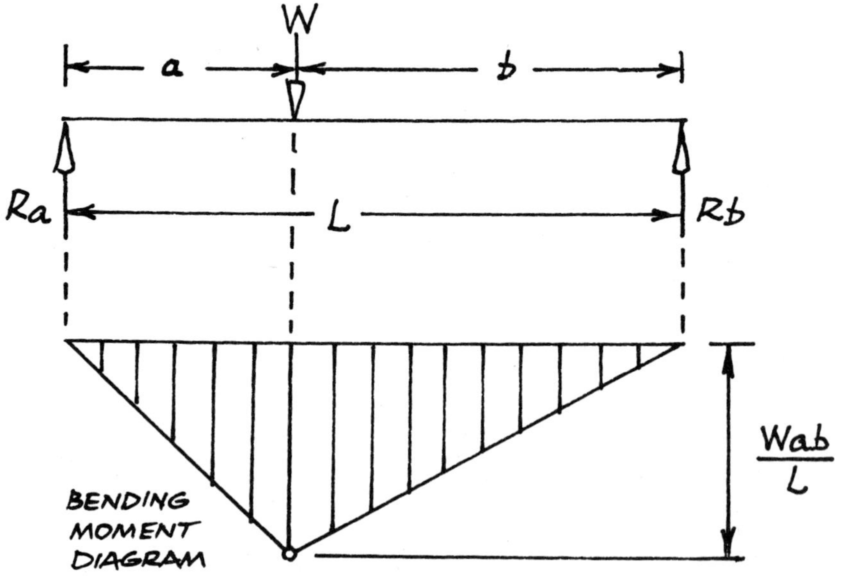

Figure 12

In Figure 12 we see the point load at a distance of "a" from one end and "b" from the other end. The first thing that happens is that the reaction forces at each end of the beam are no longer equal and so must be called Ra and Rb for the time being. However, we can see what these will be almost by inspection. For instance Ra will be greater than Rb because the load W is nearer to it. Further we know that if the load W moves all the way over to being above Ra then Ra will equal W and Rb will be zero. (The beam itself for these exercises has no weight). So in the general position Ra and Rb will each be a proportion of W and in fact it is the proportion in which the span is divided. Here then,

$$Ra = b/L \times W \text{ and } Rb = a/L \times W.$$

You can now test this by checking the known reactions when W is at the centre and at one end, i.e. when "a" is L/2 and L.

In more complicated cases we shall need to work this out. How we do this is by taking moments again but about one of the support points. We have seen that this eliminates that reaction from the equation leaving the other which is expressed in terms of W, (which is known). We also know that the bending moment at the end support point of a single span is zero and therefore here the moment in one direction minus the moment in the other direction is zero. In other words one moment equals the other and cancels it out.

Take moments then about Ra.

$$Rb \times L = W \times a$$

or

$$Rb = Wa/L \text{ as we saw above.}$$

We can now look at the general case of maximum bending moment arising from a point load anywhere on a single span beam. This will always occur at the point of action of the load. In Figure 12 if we take moments at this point of maximum bending moment which is at load W for either end of the beam we have a bending moment of

$$Ra \times a$$

which, as $Ra = Wb/L$, becomes,

$$Wab/L$$

and the bending moment diagram arising is also shown in Figure 12.

As an example let us look at a particular case of W being 600 N, L being 3000 mm and a and b being 1000 mm and 2000 mm respectively. The maximum bending moment which will be at W is,

$$\frac{600 \times 1000 \times 2000}{3000} \text{ Nmm}$$

$$= 400\,000 \text{ Nmm} = 0.4 \text{ kNm.}$$

more than one point load

We have seen that as a load or bending moment increases so do the internal forces producing the moment of resistance. As a result, if we have two point loads on a beam the bending moment diagram can be drawn by adding together the bending moment for any point arising from each. To save further pondering I set down in Figure 13 the bending moment diagram arising from the beam and loading shown.

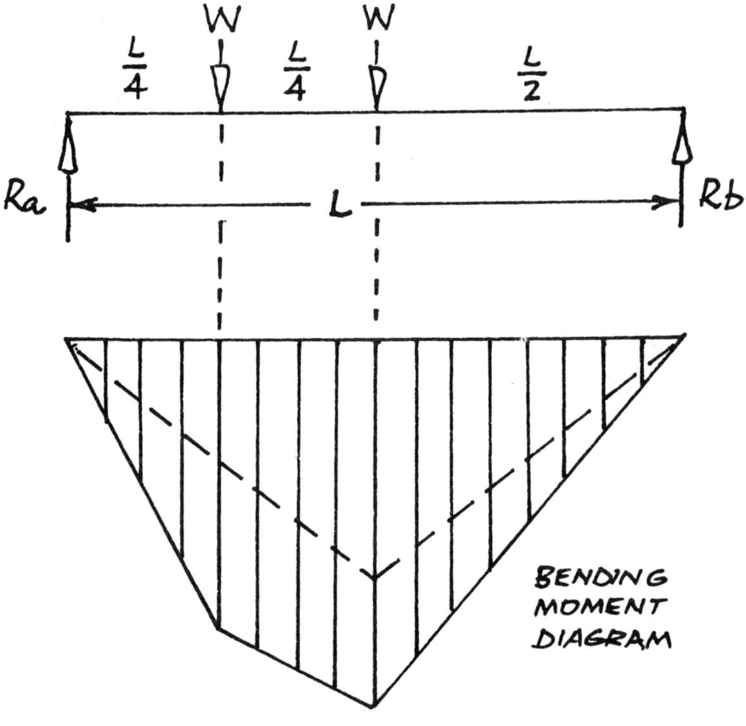

Figure 13

To obtain the values of the bending moment we first have to find out what one of the reactions is, say Rb. From the central load it is W/2 and from the left hand load, as before, it is W/4 and added together is 3W/4 (Don't forget the total load is now 2W). By the same method or by subtraction Ra equals 5W/4).

Alternatively we could have taken moments about one of the supports as before and obtained,

$$Rb \times L = WL/4 + WL/2 = 3WL/4$$

therefore,

$$Rb = 3W/4$$

To find the bending moment at the central point load, take moments about its position for the right hand half of the beam and the bending moment equals $3W/4 \times L/2 = 3/8\ WL$.

This particular answer is not worth remembering as the combinations of values and positions of multiple point loads will be many. What is important is to see **how** the calculation was done which really will be the same in all situations.

uniformly distributed load

We will now carry out exactly the same procedure for a uniformly distributed load. This is a regular load at equal intervals along the beam and as you would expect, a load per MILLIMETRE or per METRE along the beam.

Figure 14 shows a uniformly distributed load of w kN/m on a simply supported beam of length L. It also shows the usual way of expressing a "u.d." load rather than putting all the arrows in. The TOTAL LOAD here is w × the length or "wL" and as the loading on the beam is symmetrical the two support reactions will be equal therefore half the total load, i.e. wL/2.

We have seen the maximum bending moment is likely to be at mid-span and we will therefore take our moments about this point (for half the beam remember). The reason we only consider half the beam of course is that the forces in each half are turning in opposite directions. If we took the whole beam, about any point our answer would be zero and all that would prove is that the system is in equilibrium which we know anyway.

Figure 14 shows what we are doing. The turning effect of the support reaction wL/2 is in the opposite direction to that of the u.d. load (I have reverted to arrows just for this example). The moment of the support reaction about the point is therefore $wL/2 \times L/2$ which is $wL^2/4$. To calculate the countering moment of the u.d. load we could take the load on each metre or millimetre and multiply by the distance to the pivot of that particular force. That is rather involved and we get exactly the same answer if we take the total load shown on that bit of beam and use the distance of its centre of action from the pivot which is midway. We then

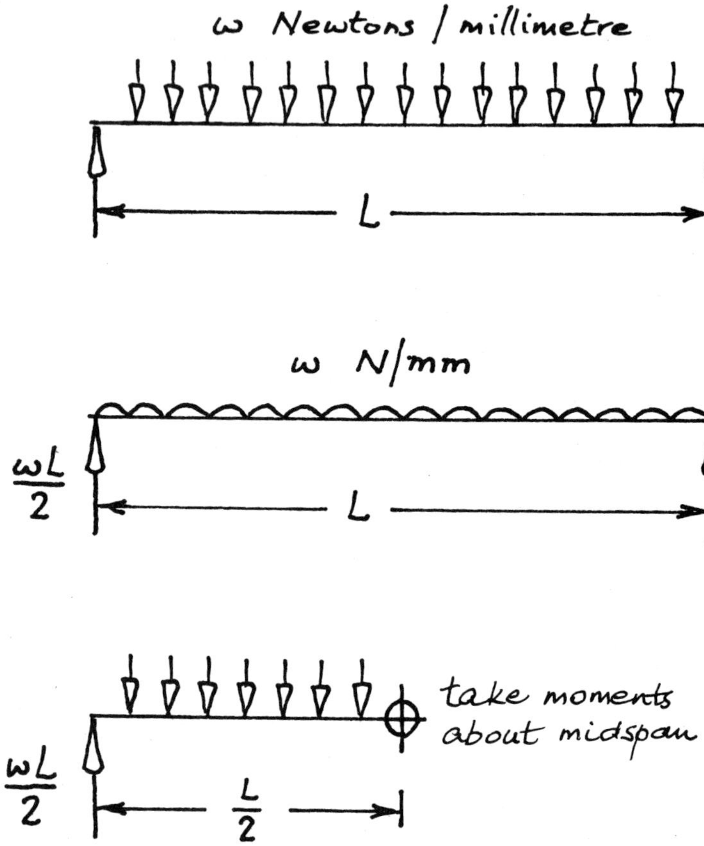

Figure 14

have the countering moment of the downward load about the pivot is therefore: $wL/2 \times \frac{1}{2}L/2 = wL^2/8$.

The nett resultant moment at midspan is therefore,

$$\frac{wL^2}{4} - \frac{wL^2}{8} = \frac{wL^2}{8}$$

This $wL^2/8$ you will use many times and is the maximum bending moment at mid-span arising from a uniformly distributed load on a simply supported single span.

For the BENDING MOMENT DIAGRAM here you could take moments at different points along the beam and plot the results. To save you the trouble the answer is shown in Figure 15.

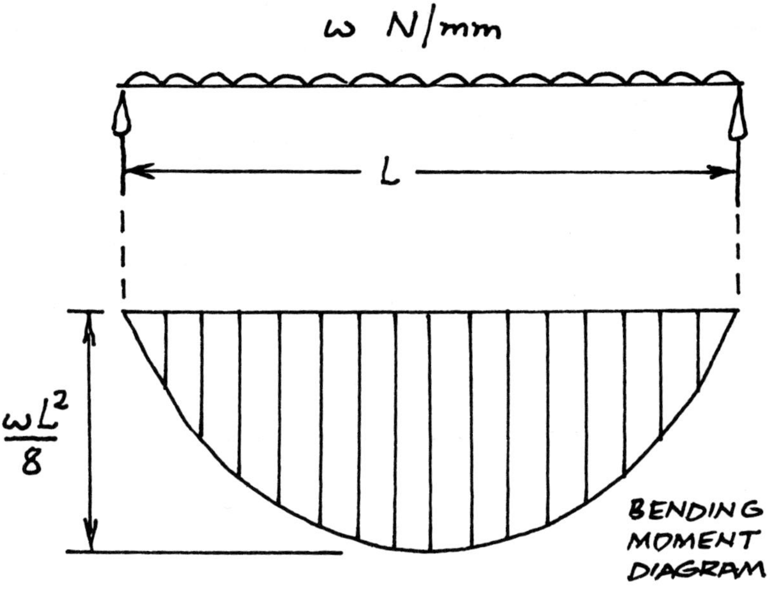

Figure 15

For the more mathematical of you, you will see that the graph has to be a curve as we have "L^2" in the answer which makes it a SECOND ORDER CURVE compared to the straight line of the single point load bending moment wL/4.

As an example in a particular case of w being 10 kN/m and L being 4·000m the maximum bending moment at mid-span is $10 \times 4 \times 4/8 = 20$ kNm.

combined u.d. and point load

We have found that there is no fundamental problem in adding bending moments arising from more than one load on the same beam. We have also been through the method for calculating bending moment at any point on a simply supported beam. This was by taking the nett moment about that point for one half of the beam. Consideration of either half will give the same answer and so we always take the half which looks simplest.

We have also seen that for a complex loading condition where we have to find out the value of the support reactions, we can take moments about one of the support points to find the value of the other. The other method was by superimposing support reactions from individual loads. So we are now armed with all the methods necessary to work out bending moment in any one position in a beam for any combination of loading.

The question of skill now arises because we shall always be wanting to find the MAXIMUM BENDING MOMENT so that we can evaluate the MAXIMUM tension and compression forces that the beam will need to withstand which will be derived from the balancing moment of resistance.

The rules in practice to find this out are simple.

1. In a case of symmetrical loading on a simply supported beam maximum bending moment will always be at mid-span.
2. In a case of a single point load on a simply supported beam mximum bending moment will always be at the position of point load, wherever it is.
3. For combinations of multiple point loads, or point load and uniformly distributed load together, maximum bending moment will be at the position of the point load nearest mid-span.

Now whilst the last rule will nearly always be true, in practice, it will obviously depend on the relative value of the individual loads if they are different. In other words a heavy point load away from centre may have a greater effect on the overall nett bending moment that a light point load at mid-span on the same beam. In any situation of doubt, and this occurs in practice very seldom, you will just have to take moments at both positions to see which is the greater. In certain very complicated cases it may not occur at a load point at all but I deal with this in the section following on shear force.

To illustrate the bending moment diagram from a combination of point load and uniformly distributed load look at Figure 16. I have shown the bending moment diagram arising from the above combination but with the off-centre point load.

You will see how one can draw a bending moment diagram as above without any recourse to calculation. Having selected the point of maximum bending moment you can proceed to calculate the value.

You will also see from this illustration that the shape arises from adding together the two independent shapes as in Figure 13.

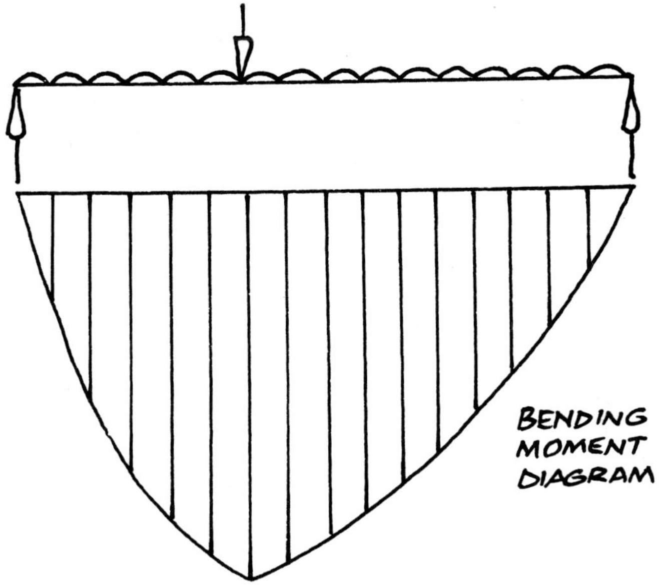

Figure 16

cantilever

Whilst most of you know what a cantilever is I will carify the matter. A cantilever is a beam with a support at one end only. There obviously has to be something more than that otherwise it would fall down. That something more is a MOMENT at the support which keeps the cantilever in place. This makes the support a FIXED END SUPPORT and the moment is called a FIXED END MOMENT. You can imagine the situation by holding a broom horizontally at one end only. Your arm muscles will soon tell you about fixed end moments in keeping the broom level.

The evaluation of cantilever moments uses the same principles as in the simply supported beam and the value of the fixed end moment is found by taking moments at the support for the loads on the beam.

This is shown in Figure 17a. In this case the fixed end moment is $W \times L = WL$. Taking a moment halfway along the cantilever would produce $W \times L/2 = WL/2$ and again we have a triangular bending moment diagram. Uniformly distributed load is dealt with in the same way as we dealt with it in the simply supported beam. The total load in

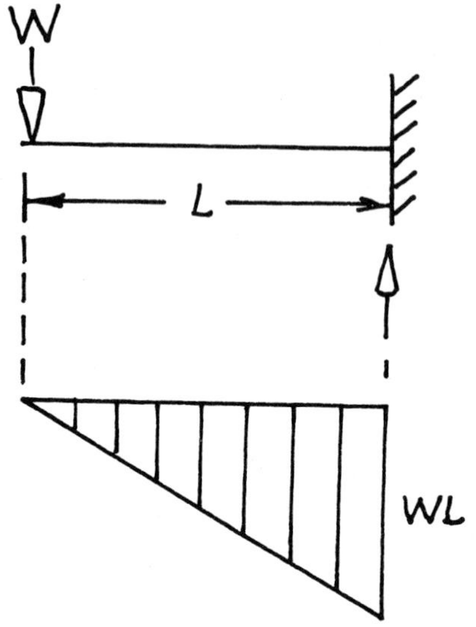

Figure 17 (a)

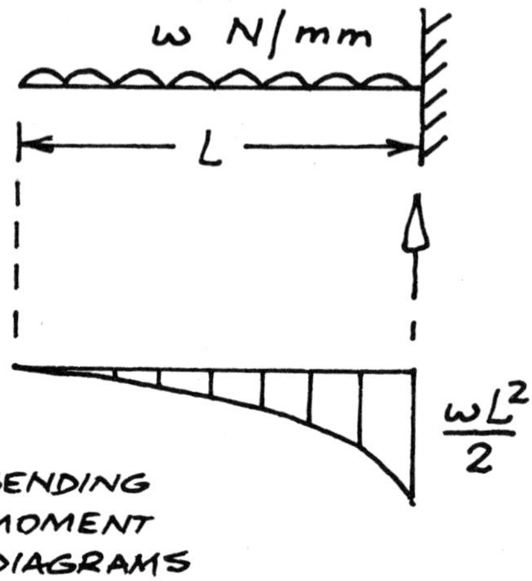

Figure 17 (b)

Figure 17b is w × L = wL and when it is multiplied by the position of its resultant line of action, the fixed end moment is wL × L/2 = $wL^2/2$. It is interesting to note that if we represent the total uniformly distributed load in this fixed end moment by W which is w × L it becomes WL/2.

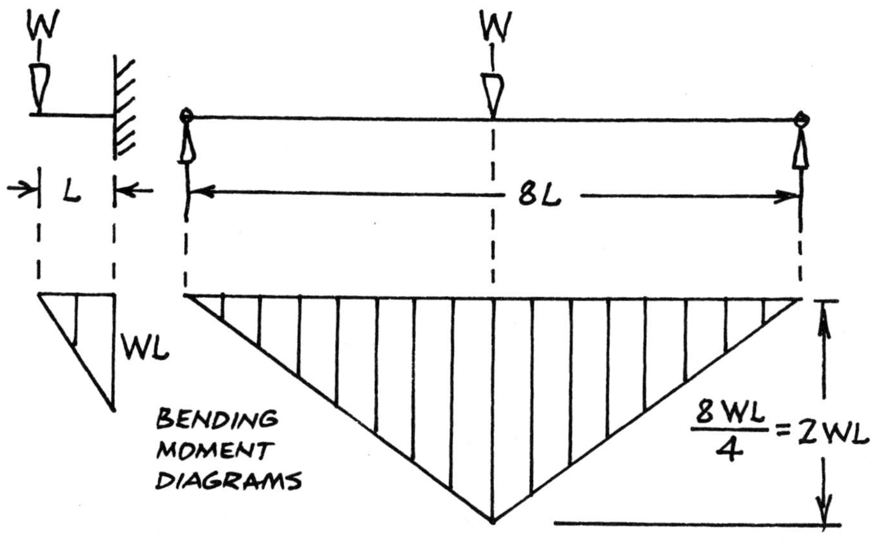

Figure 18

The conclusion therefore is that the fixed end moment for a point load at the end of the cantilever is exactly double that for the same load spread evenly over the whole beam. This also applies at mid-span of the simply supported beam.

The usual situation which we meet in practice for a cantilever is that of an overhang beyond one support joist or beam. As you would expect, any loads on that cantilever will tend to relieve the bending in the rest of the beam. We will now see how this can be easily evaluated. In Figure 18 I show a cantilever length L and separately a beam of length 8L, each with their point loads as shown and each producing their individual bending moment diagrams as shown and still in the unconnected situation.

Figure 19 shows the situation when that cantilever and its point load are an extension of that beam over the support. As bending moments cannot be discontinuous or change in amount instantaneously at a point, unlike shear force, we have to deal with what would otherwise be a discontinuity at the left hand edge support. There are various methods of considering this situation but the simplest and certainly the easiest in practice is as I have shown, to join the top of the fixed end moment in a straight line to the far support. The values of bending moment are in between these lines as shown in the shaded area.

With internal stresses you can imagine that at the fixed end of a cantilever tension is now on top and compression on the bottom. We call this

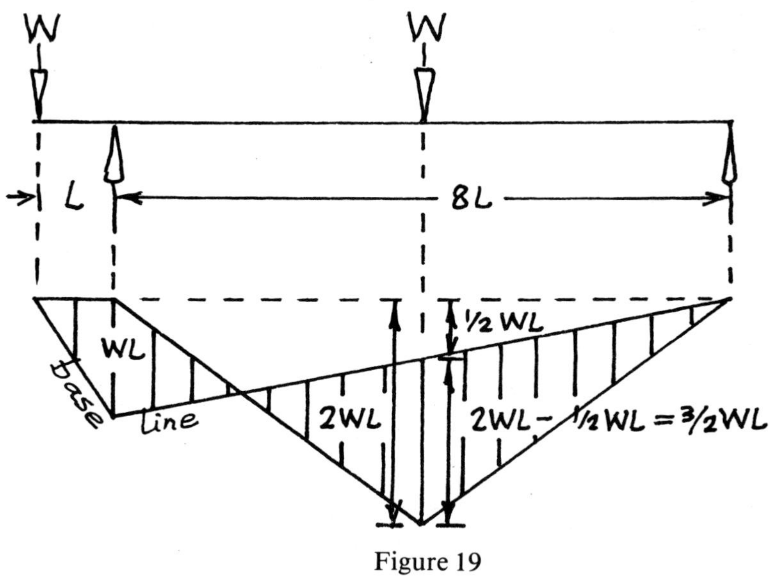

Figure 19

moment, negative bending moment, and the ordinary single span beam bending moment, positive. With the base line as shown in Figure 19 you can see how and where the negative bending moment changes to the positive bending moment. This point is called the POINT OF CONTRA-FLEXURE where the internal stresses change over. At that particular point because there is zero bending moment there are also zero bending stresses.

The principle of joining up with a straight line and producing a base line which is bent can be used generally for these simple beam extension cantilevers and it is worth trying out the situation with other loading conditions and span lengths. This will be dealt with further in the chapter on calculations.

summary

It is worth making two points of summary after this rather meaty section.

The first is that, whilst this may have seemed very complicated at the first read through, in practice you will usually be dealing with either u.d. loading by itself, a point load by itself or at worst a combination of the two. I do show a worked example later of something which has extra point loads, but only to serve as illustration and reference, should it happen to you.

The second is that for the student in an examination question you will usually be told about which point the moment is required. In the theoretical examination situation this may not always be the worst bending moment and so you will have to take these moments about the point required for one end of the beam rather than the short cut superimposition. It will also be safer to take moments about one of the reactions to find the value of the other rather than use superimposition of the individual load reactions.

shear force

You will be pleased to know that this is rather simpler than bending moment.

shear force

We first encountered shear force in Figure 9. This will vary as did bending moment as we move along the beam. However, because this is a simpler animal its variations are simpler.

Shear Force in fact is the nett cumulative vertical force of reaction arising from the imposed loads. In Figure 9, unlike bending moment we saw that because the support reaction force was constant the shear force was also constant in that part of the beam and at a value of W/2. If we had moved our "cut" point to the right however beyond the load position at mid-span, the nett vertical shear force would have been affected by the vertical force W which would have had to have been subtracted from it. The result would still have been W/2 but negative, i.e. −W/2.

Figure 20 will clarify the above and shows the reader the SHEAR FORCE DIAGRAM for a central point load as in Figure 9.

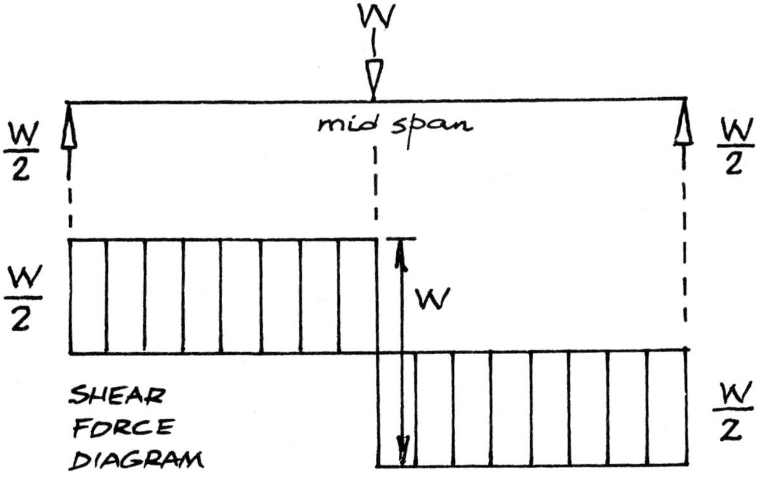

Figure 20

The value of the shear force at any point in this example is the shaded area and is constant at W/2. The change to a negative figure is immaterial from a value point of view.

This nett vertical evaluation of shear force will become more apparent in the uniformly distributed loading case. In this case as we move from left to right we find that whilst the support reaction remains constant at wL/2, the **nett** vertical force is being steadily reduced, for each millimetre we progress, by the gradual inclusion of the uniformly distributed load in the other downward direction. When we get to mid-span the **nett** shear force has disappeared altogether only to build up again as we approach the right hand support (but again here with the negative sign).

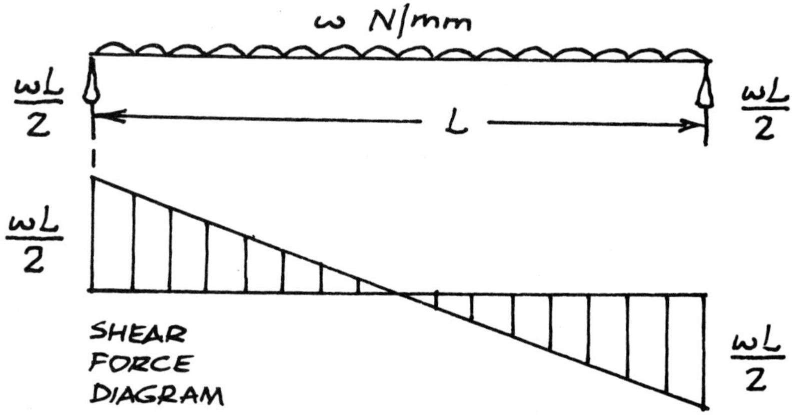

Figure 21

Figure 21 shows this SHEAR FORCE DIAGRAM which we see is of triangular shape.

To draw a shear force diagram for a simply supported beam then we start from the value of one of the support reactions and work our way across subtracting the downward loads encountered until we reach the other side and end up with the value of the reaction at that point. MAXIMUM SHEAR FORCE for a simply supported beam is always at and is the same value as the larger of the two reactions (or at each reaction if the loading is symmetrical).

Superimposition also works here as before and the shear force diagram for the loading from Figures 20 and 21 acting together is now shown as Figure 22.

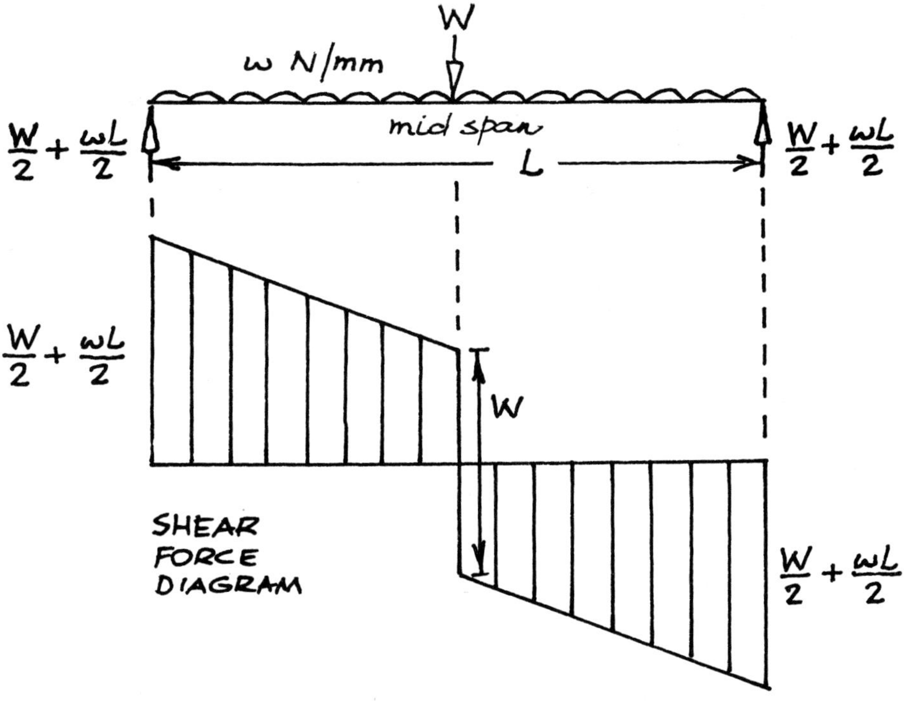

Figure 22

shear force and bending moment

If you look at all the illustrations we have used in this chapter two rather significant points emerge.

1. Maximum bending moment occurs at the same point as zero shear.
2. Maximum shear force occurs at the same point as zero bending moment.

These two points are of little use in the simply supported beam which is largely the scope of this book but the relationship is significant.

The point of zero shear can be of use as referred to when we were

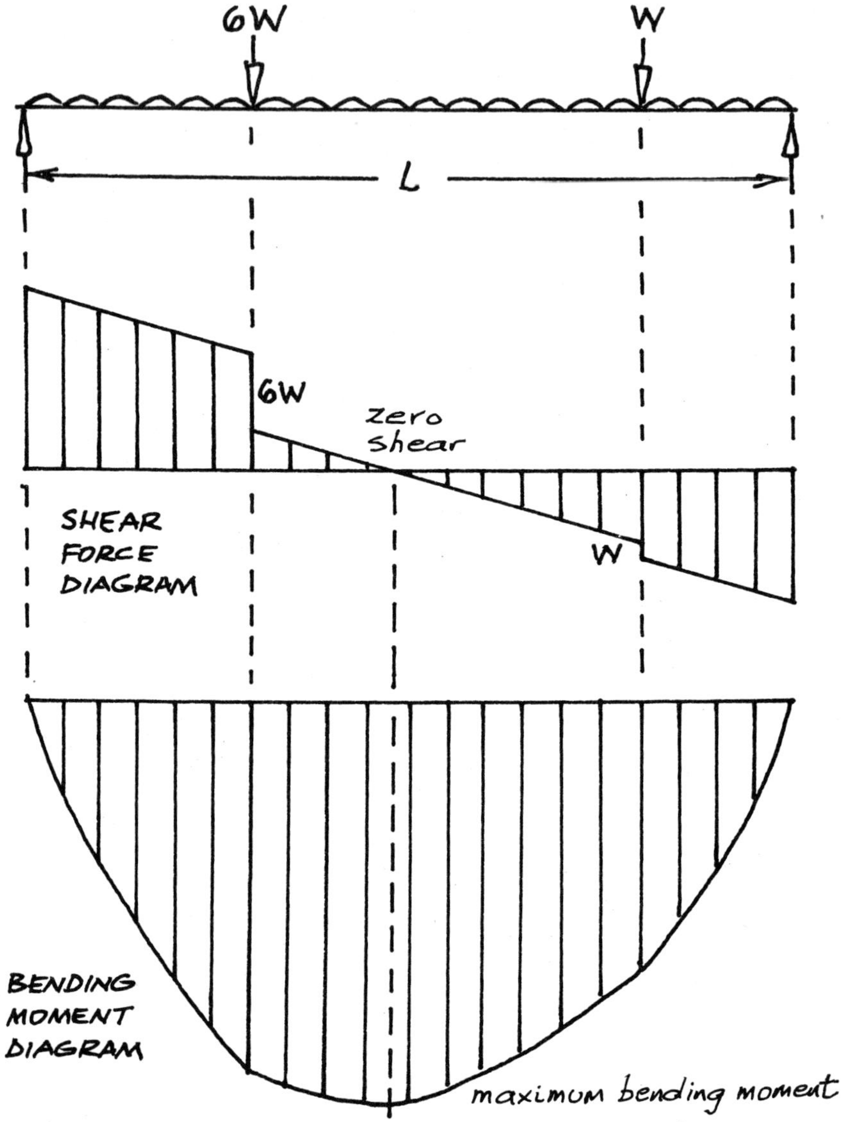

6W
W
L
6W
zero shear
W
SHEAR FORCE DIAGRAM
BENDING MOMENT DIAGRAM
maximum bending moment

Figure 23

trying to find out the precise position of maximum bending moment. As I said, in some situations it might not occur exactly at a loading point. In these cases and where we have to be precise we can use the above the relationship. This is done fairly easily by finding out the point in the beam at which the downward loads on that side become equal to the support reaction on that side. This complication only occurs when we have u.d. load as well as point loads and I illustrate the point diagrammatically in Figure 23 by drawing bending moment and shear force diagrams for the same loading conditions.

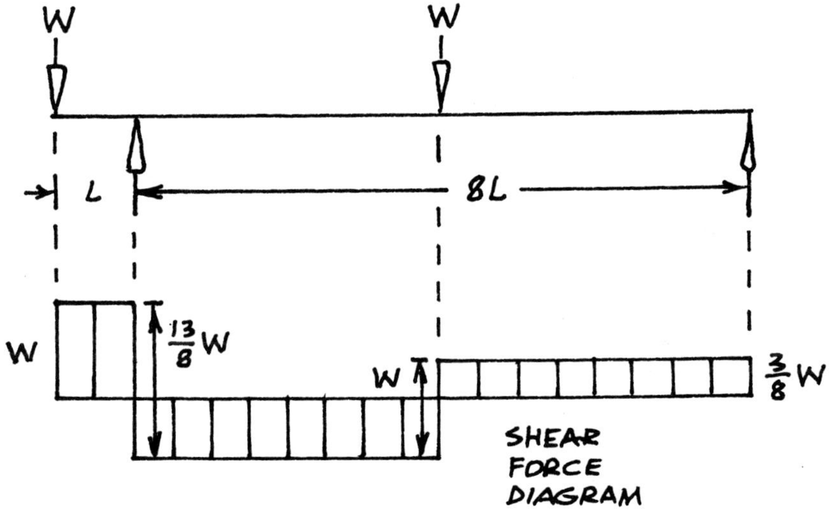

Figure 24

You will see that this is the precise position of maximum bending moment. For **practical** use, however, you will also see that our previous rule locating it at the point load or mid-span would not have been far out either in position or value.

Figure 24 shows the shear force diagram for the cantilever beam extension from Figure 19. To do this we need to know the reactions Ra and Rb and we do this as before by taking moments about one support, say Rb:

$$9WL + 4WL = 8RaL,$$

therefore, $Ra = 13W/8$ and $Rb = 2W - 13W/8 = 3W/8$.

conclusion

At this stage we have looked in principle at where loads come from and how they can be calculated for a particular position. We have looked at how our beam "works" and can see how the maximum bending moment and shear force can be calculated in a particular situation.

We shall be looking next at how different materials resist these internal tension, compression and shear forces which may prove more interesting.

Chapter Three

The supporting cast

We now know that there are forces in all load-bearing parts of a building. These are either beams, as horizontal elements, or walls, as vertical elements. We also know how to calculate what those forces are.

We will now look at these supporting members and see how the forces acting in them are resisted by the various materials.

Up until now we have been using FORCES in our examination of what have been largely effects caused by external situations. Now that we are going "inside" the materials you will see that the animal we shall use mostly is STRESS. You will recall the relationship from Chapter One but if you don't, look back again.

We first look at beams and spend most of the time seeing how those forces of tension and compression which act together to form the MOMENT OF RESISTANCE are distributed throughout the depth of the beam. First we will remind ourselves of the situation by enlarging our beam for clarity and exaggerating the deflection which would be sustained by the loading.

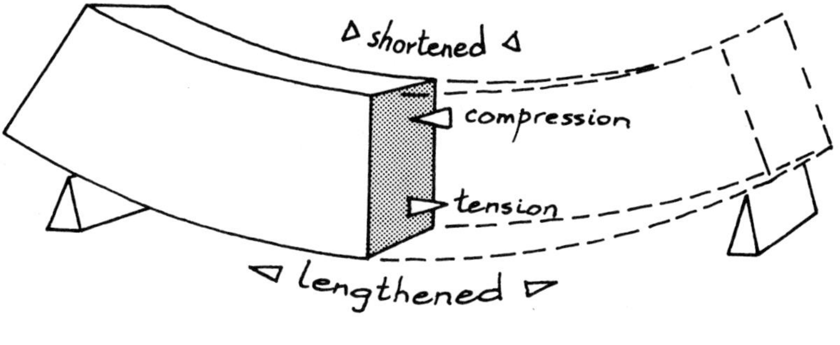

Figure 25

Figure 25 shows this enlargement and we can see DEFLECTION with which we are familiar. A minute or two of analysis of this common phenomenon may be useful.

You will recall that deflection is a form of STRAIN and that strain is the deformation of material under STRESS. From the exaggerated deflection in Figure 25 we can see that the top surface of the beam has **shortened** under this strain and the bottom surface has **lengthened.** So here straight away is confirmation of our earlier theoretical analysis that the STRAIN in the top half is caused by "squashing" COMPRESSION forces and that in the bottom half by "stretching" TENSION forces. We can also see that the amount of shortening and lengthening gradually changes downwards across the beam. This changes from maximum shortening or COMPRESSION on the top to maximum lengthening or TENSION at the bottom. Where these change over halfway down there is zero strain and therefore zero stress. This position is called the NEUTRAL AXIS.

These are physical characteristics in any beam in bending but as different materials have different characteristics we shall now see how they each behave. We shall then move very quickly to actual evaluation of **strength** which then, at last, can match up to the load we are applying.

the timber joist

We have seen that a simply supported beam under load suffers maximum compression in the top fibres, zero stress on the neutral axis and maximum tension on the bottom fibres. This variation down through the beam is of a straight line nature. In other words, half way from the top of the neutral axis the compressive stress is half the value at the top. We can now draw a graph of this (Engineers love drawing graphs. In our defence, this sort of diagrammatic representation is very helpful and I therefore make no apologies!).

Figure 26 shows a STRESS DIAGRAM across our beam. We see our old friends the compression and tension forces, C & T and we see the STRESS that these are translated into across the internal plane shown. We have seen why this STRESS is not the same all over (unlike direct compression or tension). This triangular variation is because of DEFLECTION and BENDING and is therefore called BENDING STRESS or sometimes FLEXURAL STRESS. This name tells us that we have tension and compression acting together in opposite directions.

Remember that STRESS is the amount of FORCE per unit area (usually millimetres). In Figure 26 we see that the small arrows represent-

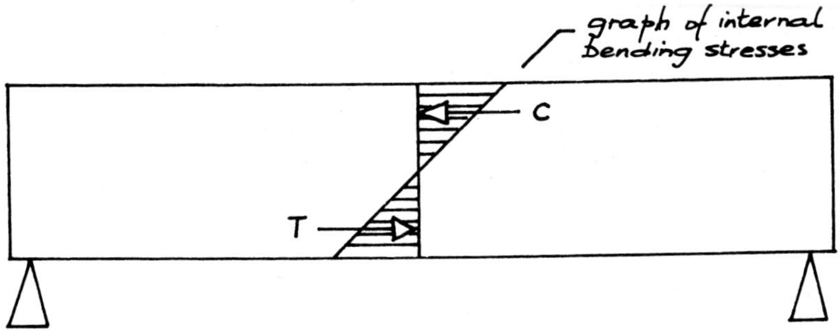

Figure 26

ing STRESS reduce in scale length and therefore amount as we move down the beam to the neutral AXIS and then increase again in the other direction (now representing tension) to a maximum at the bottom. The addition of all the compressive FORCES PER SQUARE MILLIMETRE over the area on which they are acting will produce a TOTAL COMPRESSION FORCE – and the same for the tension half.

Now because the nett compression FORCE is a collection of all the individual STRESSES, its line of action will be at the "centre of gravity" of the triangle of stresses. As you know this is ⅔ up from the sharp end of any triangle. The same happens here and also with the tension half. We can now therefore see that the distance between the NETT COMPRESSION FORCE and the NETT TENSION FORCE is itself ⅔ of the OVERALL DEPTH of the beam. You will remember that this distance between forces acting equal, opposite and parallel was one of two factors affecting the value of this COUPLE or MOMENT OF RESISTANCE.

The MOMENT OF RESISTANCE is therefore,

$$\text{Force C (or T)} \times \tfrac{2}{3}\,d.$$

This does not help us a great deal as the strength of different timbers or of any material is determined by a MAXIMUM DESIGN STRESS (f) which is descriptive of the material in general rather than what size a particular piece might be. So we have to express our Force C back into stress terms for the particular beam we are examining.

The nett force C (or T) will be the average stress × area of surface over which it is acting, i.e. $f/2 \times (b \times d/2)$.

This is the "volume" of the "wedge" of compression stresses in Figure 27.

The MOMENT OF RESISTANCE ($C \times \tfrac{2}{3}\,d$) now becomes,

$f/2 \times (b \times d/2) \times \tfrac{2}{3}\,d = f \times bd^2/6$,
i.e. Moment of Resistance $= f \times bd^2/6$.

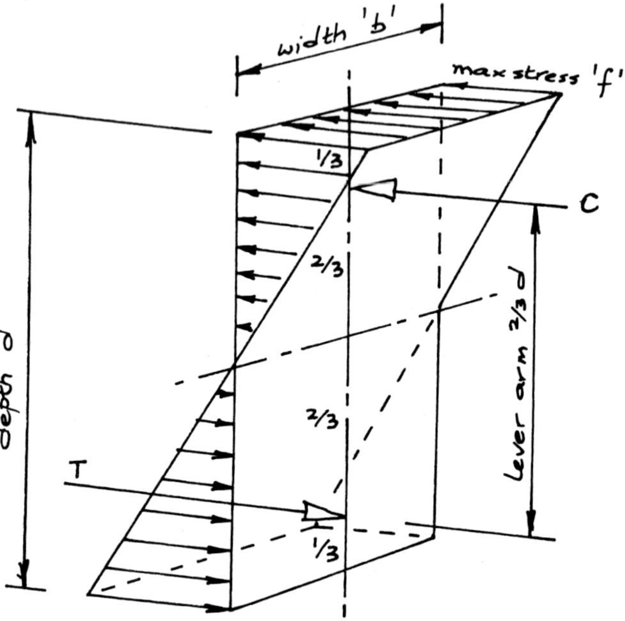

Figure 27

From a known beam size, "b" and "d" will be known, and with an allowable maximum design stress "f" we can work out the maximum moment of resistance.

Look at the above moment of resistance. You can see it is the allowable maximum design stress × figures derived from the shape of the cross-section. These figures (which for all rectangular homogenous beams is $bd^2/6$) are called collectively the MODULUS or SECTION MODULUS. The units of MODULUS by inspection of $bd^2/6$ are therefore mm^3. This is represented by "Z" and becomes more relevant in the steel beam later. However, the principle remains the same, that is,

$$\text{that, } M = fZ.$$

This equation is central to all calculations for bending strength of beams and in proper use will gradually become second nature to you.

The principle of all beam design which we can now understand is that Moment of Resistance has to be equal to or greater than the maximum applied Bending Moment.

Let us do a simple example to help register the point in your mind.

Normal clean, straight, building carpentry timber would these days normally be classified as GRADE 50. This has an allowable maximum DESIGN STRESS in compression and tension of $5 \cdot 0$ N/mm².

Question

Find out the moment of resistance of a floor joist 200mm × 100mm of grade 50 quality.

Answer

Moment of Resistance $= f \times bd^2/6$

$$= \frac{5 \times 100 \times 200 \times 200}{6} = 3\ 300\ 000 \text{ Nmm} = 3.3 \text{ kNm}$$

This we can see would be unnecessarily strong for the loading example on page 25 but nowhere near strong enough for the example on page 29.

So here we can get an indication of the balancing operation of the design process. In the Appendix, I have provided Tables of moments of resistance of timber beams and others which will considerably speed your design process. However, you must know how these are worked out both for the good of your soul and so that you can calculate a beam size not included.

the steel beam

Steel is both stronger, heavier and more expensive than timber, or concrete, and therefore it is never used in a solid rectangular form. The traditional "I" shape of the rolled steel joist has come about as follows. Due to these characteristics of steel, it has to be shaped in a beam so that as much of it as possible is stressed to the maximum. Therefore providing steel to resist all the smaller bending stresses near the neutral axis as in Figure 27 is inefficient and uneconomic. As much steel as possible is placed nearest the top and bottom to resist tension and compression at their maximum and with the distance between them also at a maximum. The "web" or vertical part in the middle keeps the top and bottom "flanges" apart and by doing so is resisting SHEAR. The top and bottom flanges resist the compression and tension and the web resists the shear.

Figure 28 shows a steel joist. The slopes of the flanges and radii at junctions arise from both requirements of the rolling manufacture

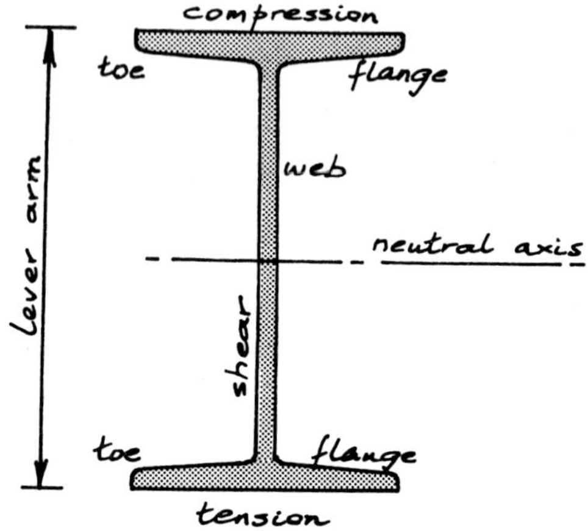

Figure 28

process and to save local stress concentrations between flanges and web which would arise if everything was square.

As steel, like timber, has fairly equal strength in tension and compression the **principles** of moment of resistance are as in the timber joist. For moment of resistance the reduced stresses of compression and tension which there will be down the web can be disregarded. This is because their reduced size nearer the neutral axis and their reduced distance apart, mean that their moment effect will be negligible. By the same logic the tension and compression stresses over the flanges will be considered uniform, because the actual variation over the shallow depth will have negligible effect on the moment value.

In such a shape as this the MODULUS would be very difficult to calculate as we did with the rectangular timber beam. The steel manufacturing industry has therefore done it for us. Each steel joist and channel has been given a MODULUS in units of cm^3. (For most of our work we have to convert this to mm^3 by multiplying by 10^3 or 1000). I have reproduced those relevant to most of our work in the Appendix. The principle of this MODULUS is the same as for the timber beam. It is a figure representing the area of a flange multiplied by the distance between flanges. Therefore when this MODULUS is multiplied by the DESIGN STRESS we have the MOMENT OF RESISTANCE.

Again let me illustrate with an example.

A 127×114 steel joist developing its full compressive (and tensile) stress of 165 N/mm^2 (compare this to timber) will have a moment of resistance as follows:

$M = f\,Z$
$M = 165 \times 148{\cdot}8$ (Z from the table in cm^3) $\times 10^3$ N mm
$M = 24\,552\,000$ Nmm

which approximately equals 24·5 kNm.

Here we can see a more appropriate strength figure to resist the bending moment caused by the loads on page 29.

adequate lateral restraint

Before going any further we must deal with one very important general point applicable to all beams, but has particular significance in use for the steel beam.

DESIGN STRESSES have been developed by the British Standards Institution for all materials. These design stresses are based on FAILURE STRESS, i.e. the stress at which physical failure of the material takes place, e.g. the breaking of a piece of string in tension. A factor of safety has then been introduced of either two or three for different materials so that the WORKING or DESIGN STRESS in the material under expected maximum loads is SAFE. The SAFE WORKING STRESS for a material as decided by the British Standards Institution is therefore our DESIGN STRESS, whether for timber, steel or concrete.

The example of our piece of string in tension is easy enough.

Compression failure is more complicated because, to develop its full compression strength, a beam (or column) in whatever material will need to be laterally braced or held, otherwise it will buckle sideways, long before the maximum internal compression stress is reached. As an example consider the extreme case of a 100mm × 50mm timber, 15 metres long (if one could be obtained). Supported at its ends this timber would twist over sideways even under its own weight. From a stability point of view therefore, that beam can be considered to have FAILED even though the compression stress developed is well below the compression failure value.

Whilst this applies equally to a timber as any other, I did not complicate that section with this matter as almost without exception a timber joist will have floor boards nailed to the top which quite adequately brace the compressive top half against buckling. In other situations as well, as in roof rafters bracing the top of a purlin, or providing herring-bone strutting in a long span floor, timber beams are nearly always naturally braced in the proper building construction process.

Steel is quite another matter and in domestic or commercial building one can almost say that the opposite is more often the case. The reason for this is that steel is a less easy material to fix to and generally any bracing provision has to be fixed to the steel to ensure that it does its job for at least the life of the building. I discuss later what methods can and cannot be considered as bracing a steel beam but let us look now at how we get over the problem when we cannot adequately brace our steel beam.

The design policy adopted by the British Standards Institution to offset the chance of compression buckling is to reduce the ALLOWABLE DESIGN STRESS below the normal for maximum compression. These reduced design stresses are set out in a table in the Appendix. As the beam becomes more slender the recommended design stress further reduces to offset the otherwise greater chance of compression buckling.

The factors in a particular beam which govern slenderness may seem initially rather complicated but they are simply figures which represent the cross sectional size of the beam and its length. So for each beam in its location we have to evaluate two special figures or characteristics about the beam so that we can look up the allowance design stress in those circumstances.

The first is beam depth (D) divided by the mean thickness of the flange (T). This D/T is given to us ready worked out for each joist and channel section in the steel tables, so that's easy.

The second will vary according to the location because it uses the beam length (l). This length (in mm) has to be divided by another characteristic of the beam section which we can look up and that is called the radius of gyration (ry). This is given to us in cm in the tables. As we usually work in mm rather than cm., we have to remember to have the units top and bottom of this ratio the same, 1/10ry.

Let us look again at my example on page 47 and assume that it is 4·000 metres long and that it does not have adequate lateral restraint.

From the tables, D/T for this beam is 11·0.
1/10 ry in this case is $4000/10 \times 2{\cdot}55 = 160$.
From the reduced stress table in Appendix VII the design stress is now reduced from 165 N/mm² to 158N/m m².

The new moment of resistance in this unbraced

situation is now,

$$M = 158 \times 148 \cdot 8 \times 10^3 = 23\ 510\ 000 \text{ N mm.}$$
$$\text{or } 23 \cdot 5 \text{ kNm.}$$

In this particular case the beam is still strong enough for the bending moment on page 29 but you can see how the reduced ALLOWABLE DESIGN STRESS has reduced the allowable MOMENT OF RESISTANCE compared to that on page 47. In other situations the reductions can be considerably more as you can see from the range of reduced stresses in the table. In some cases you might have to select a bigger joist size in circumstances without lateral restraint to achieve an allowable moment of resistance large enough to balance the applied bending moment.

You will see more of this in use in the examples later on and really it is a lot less complicated than it looks here.

the concrete beam

Concrete is funny stuff.

To start with it is the only structural material in building which is not found as such in nature. Timber is directly from nature. Steel and iron is pretty well from nature even though it is smelted and shaped. Brick is also from nature – again fired and shaped. Concrete is a collection of natural materials (sand and gravel) bound hard together in the cold state by cement which itself is manufactured through natural lime.

It is the newest of our basic building materials and for the last 100 years we have learnt how useful it is. Its main characteristics are as follows: 1. hardness and strength quite good (between that of timber and steel); 2. because it is formed in the wet state it can be made into any shape we like; 3. it is relatively cheap (again between that of timber and steel). A summary of the above points is that it can do jobs in the building industry that other materials cannot do, e.g. slabs both on the ground and suspended and beams to any shape and size.

The main difference in its characteristics compared to timber and steel is that, unlike these other two materials, it does not have equal strength in compression and tension. This difference in strength is to the extent of it being at least 15 times stronger in compression than tension. This great difference means that from the structural point of view it has **no strength in tension**.

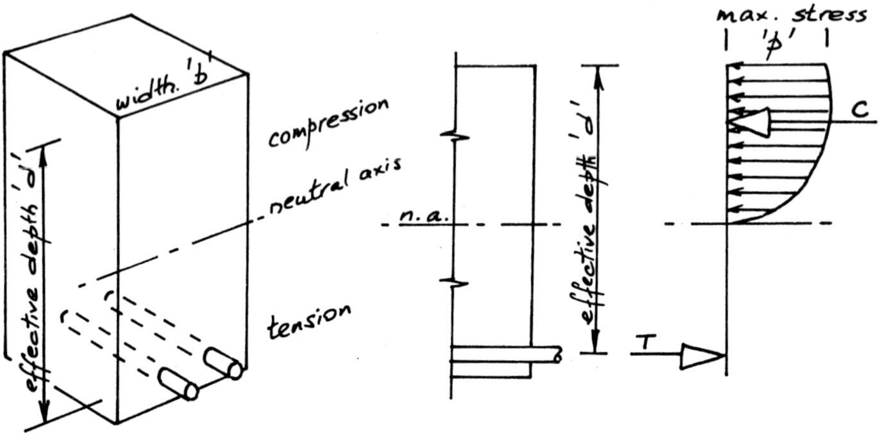

Figure 29

You will recall that in any beam there is a nett tension and compression force caused by bending. Because concrete has no strength in tension we have to provide something in the bottom half that can carry the tension force and balance the compression in the top half. For this we use steel rods around which the concrete is cast and on to which it bonds. These steel rods are **reinforcing** the concrete where it is weak in tension and the whole thing is therefore called REINFORCED CONCRETE.

A reinforced concrete beam is shown in section in Figure 29. Compare this with our timber beam in Figures 25 and 26. The similarities of compression and tension forces and neutral axis should be noticed. The main difference is that the tension force is carried wholly by the reinforcement bars.

The other slight difference which the hawkeyes amongst you will have seen is that the distribution of compression stress is not quite triangular but curved.

This is another reason why concrete is funny stuff. Without going into too much detail the stress/strain relationship in concrete is not constant as it is in timber and steel and so its behaviour under bending is not the triangular straight line distribution of timber. I mention this only to show why we do not calculate the moment of resistance of concrete in quite the same way as timber.

This curved compressive stress distribution in Figure 29 has two effects upon the moment of resistance. Firstly, the total amount of compressive STRESS which makes up the nett compressive FORCE is greater and secondly the nett compression FORCE is acting closer to the neutral axis.

Research into the behaviour of reinforced concrete has however allowed a formula to be derived which enables us to calculate the compressive moment of resistance of a concrete beam just as easily as timber.

This is $M = p \times bd_1^2/4$

Where p $=$ design compressive stress of the concrete.
b $=$ width of the beam.
d_1 $=$ effective depth of the beam.

EFFECTIVE DEPTH is the distance from the top of the concrete beam down to the centre of the tension reinforcement bars and therefore is slightly less than the **actual** depth of the beam. This is because we do not consider concrete in tension in bending strength and therefore from this point of view it is not there. Accordingly the overall depth of concrete has no relevance to moment of resistance. The amount of concrete underneath the reinforcement is usually 15mm for internal beams. Therefore by the time we have come up a bit further to the actual centre of the reinforcement bar the **effective** depth will usually be 25mm less than the **actual** depth of the beam.

DESIGN COMPRESSION STRENGTH of concrete, like all materials, has to be less than its FAILURE or CRUSHING STRENGTH. For concrete the design stress (p) is ⅓ of the crushing strength, i.e. a factor of safety of 3.

Enough of theory for the moment. Let us calculate the compression moment of resistance of a reinforced concrete beam 100mm × 200mm deep and made from concrete of 21·0 N/mm² crushing strength.

$$\text{Compression m.r.} = p \times bd_1^2/4$$
$$= \frac{21}{3} \; \frac{100 \times 175 \times 175}{4}$$
$$= 5\ 359\ 375 \text{ Nmm}$$
$$\text{or more simply} = 5{\cdot}35 \text{ kNm.}$$

Compare this now with the timber joist on page 45 which was of the same nominal size.

We have, of course, now only dealt with the top compression half of the reinforced concrete beam. This is always done first in a calculation so that we can check that the size of the beam selected is strong enough in compression (moment of resistance) for the loads being applied (bending moment). Once we have done this we can work out how much steel will be required to carry the tension force in the bottom half.

We do this by matching, this time, TENSION moment of resistance to the applied bending moment. Quite simply this is as usual Tension force × distance to centre of Compression force (lever arm). Tension force will be the cross section area of steel bars to be used (A) × allowable steel design stress mild steel (138 N/mm^2),

i.e. tension m.r. = f × A × la.

The lever arm in reinforced concrete is not straight forward because the position of the nett compression force and neutral axis varies and will be nearer the neutral axis the more heavily it is stressed. For normal use we have to simplify this situation. There are various methods and tables used to establish the lever arm in reinforced concrete and therefore the area of steel to be used. For the purposes of this book and size of reinforced concrete structures likely to be encountered by its user I am recommending the constant use of ⅔ d_1, as the lever arm. This will give a conservative estimate of tension moment of resistance and steel area.

Having established the lever arm the only unknown in our tension moment of resistance equation is A, the area of steel required for balance. Transposing the above equation we achieve A = M/f la.

In the above example of the 200 × 100 reinforced concrete beam, the area of steel reinforcement needed for an applied moment of say 1·5 kNm, would be:

$$A = \frac{1\,500\,000}{138 \times \frac{2}{3} \times 175} = 93 \text{ mm}^2$$

From the table in Appendix IX you will see that a single 10mm bar is $78mm^2$ and a single 12mm bar is $113mm^2$ and so the single 12mm is the one to select.

The above situation of having the concrete moment of resistance stronger than the tension moment of resistance is not unusual in reinforced concrete and in fact is to be preferred. In practice there will be other architectural factors controlling the dimensions of the reinforced concrete beam. The design procedure is to check that the concrete moment of resistance is big enough and then to calculate the steel reinforcement from the external bending moment actually being experienced.

At this stage we will now compare in Figure 30 the three beams we have considered so far in timber, steel and reinforced concrete. They have all been of similar external dimensions and so you will get an idea of the relative strengths of the materials.

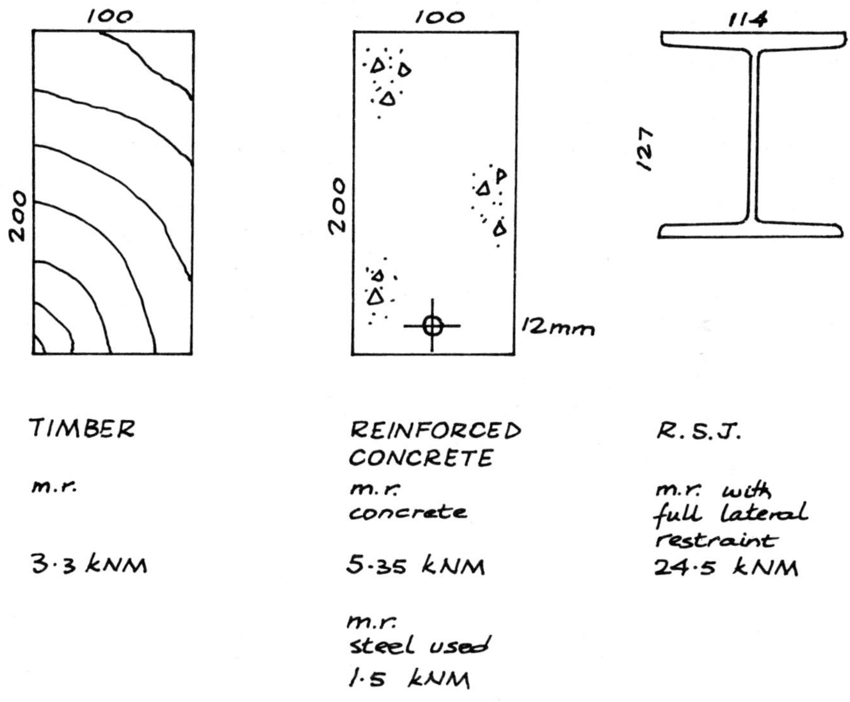

Figure 30

the brick wall

Brickwork supports vertical compression forces.

This is a straightforward section as a result. The Code of Practice for loadbearing walls B.S.C.P. III is well set out for you to understand. It will be useful however to take you through the principles with probably a little more explanation. Whilst I shall pull out the main features I shall assume that you have access to the Code of Practice for any more detailed coverage.

We are usually in the position in practice of checking whether a particular wall is strong enough for the compression loads being imposed. The principles here are no more complicated than the principles of force and stress illustrated in Figure 3. So if we know the load on a metre run of wall and its thickness, we can work out the stress sustained by dividing the load by the area of cross-section of a metre length of the wall.

For illustration we might have a length of 100mm thick solid brick wall being loaded by a force of 40 kN/m. Find the stress sustained in N/mm^3.

$$\text{The stress, } f = \frac{40\,000}{1000 \times 100} = 0{\cdot}4 \ N/mm^2.$$

The allowable DESIGN STRESS of brickwork is a bit more complicated than other materials because there are more variables in manufacture. We have to take into account both the strength of the particular

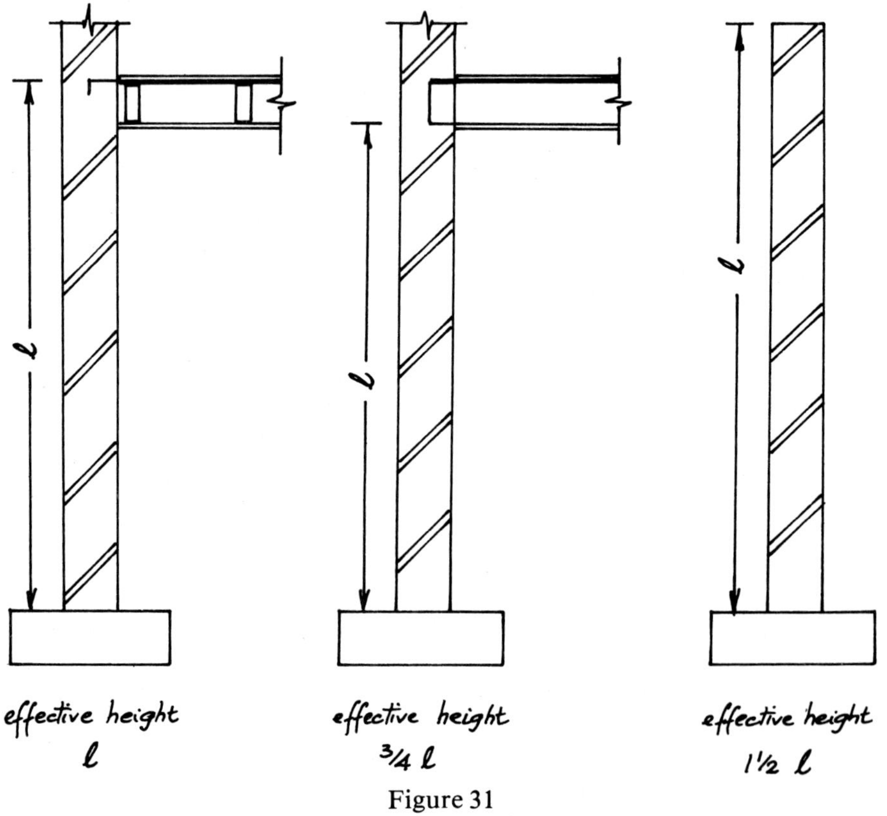

Figure 31

bricks, and of the mortar used, apart from the slenderness aspect that we have met before. We have to work first from a **basic design stress**. This is read off from the table for the particular brick strength and mortar mix which we are using.

As with the compression flange of a steel beam, a brick wall will buckle and collapse long before the stress for pure compression crushing failure has been reached. Similarly, the design principle of reducing the design stress so that buckling failure will not occur also applies to brick walls.

Slenderness for brickwork is easier than with a steel beam. The SLENDERNESS RATIO of any wall is height divided by thickness. However, different types of wall construction affect the slenderness and we have to use the terms EFFECTIVE HEIGHT and EFFECTIVE THICKNESS.

EFFECTIVE HEIGHT is governed by the end conditions of the wall top and bottom. If the wall has a floor running parallel and tied to it at the top but not bearing on to it then the effective height is the actual height between the foundation and the floor above. If the floor (or roof) bears into the wall extra stiffness is being provided and the effective height is ¾ × the actual height. On the other hand if the wall has neither of these and is free standing the effective height is 1½ times the actual height. This is obviously rather rare for it to be carrying load. These are illustrated in Figure 31.

EFFECTIVE THICKNESS for a solid or bonded wall is the actual thickness. For walls not solidly bonded or cavity walls the effective thickness is ⅔ the actual thickness of the skins of brickwork or blockwork. These are illustrated in Figure 32.

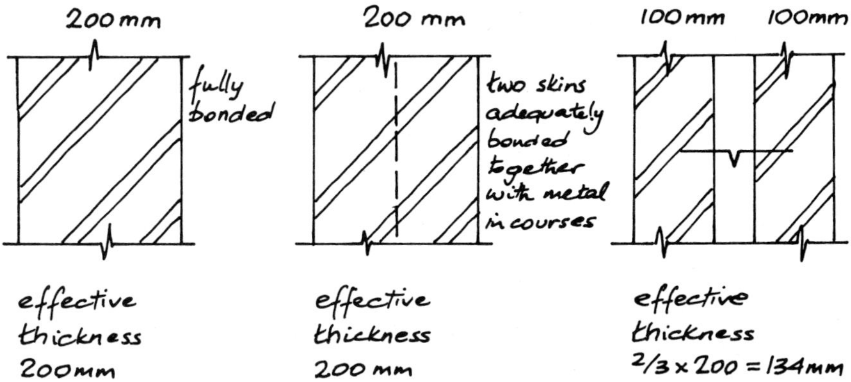

Figure 32

When we have our slenderness ratio for the wall in question, we can read off the STRESS REDUCTION FACTOR to be applied to our BASIC DESIGN STRESS to find out the PERMISSIBLE COMPRESSIVE STRESS.

So now we are in business.

Let us find out the maximum load per metre run that can be applied to a 250 mm cavity wall in common brickwork of height 3·000 m between foundation and a first floor parallel to it.

The crushing strength of common bricks can be assumed at 20·5 MN/m^2 and the basic stress from Appendix XIV for units with crushing strength of 20·5 built in 1:1:6 mortar is 1·30 MN/m^2 (or N/mm^2).

$$\text{Slenderness ratio} = \frac{3000}{2/3 \times 200} = 22$$

Stress reduction factor (from Appendix XV) for a slenderness ratio of 22 = 0·56.

Permissible Compressive Stress = 0·56 × 1·30
= 0·73 N/mm^2

Permissible load per metre run
= 0·73 × 1000 × 200 = 146 000 N
= 146 kN
= 14 600 kg Mass

In summary the method for all load bearing walls is

(1) slenderness ratio = $\dfrac{\text{effective height}}{\text{effective thickness}}$
(2) find stress reduction factor from table for slenderness ratios, Appendix XV
(3) basic stress from table for the particular wall construction and crushing strength Appendix XIV
(4) permissible working stress = stress reduction factor × basic stress

The Code of Practice also covers the more normal situation for a cavity wall of having materials of different strengths in each leaf as with an inner blockwork skin. The answer is that for the slenderness ratio calculation everything is as before. However, the Basic Stress has to be the **weaker** of the two materials. Apart from safety, this is also often logical as the weaker inner blockwork skin is usually the one carrying the load. It is however acknowledged by the Code that the outer skin is contributing stiffness and therefore reducing slenderness.

point loads

The Code of Practice also allows an increase in Basic Stress of 25% where point loads are involved in a wall, e.g. bearing for a truss or steel joist. This can apply to the area underneath a padstone. The reason is that the normal 45° distribution within the brickwork will spread the load enough within millimetres below the padstone to allow a slightly heavier stress to start with. This is of course only a local increase in **design** stress as distinct from crushing stress.

the column

The column in domestic and small commercial construction is like the brick wall in that it only supports vertical compression forces. In this way it acts as a prop to a beam often at one end. This is something at a corner allowing pedestrian passage either side. Also like the wall, design is based on and controlled by its liability to buckle. In this respect, however, the material used makes a difference in detail but not in principle. This difference derives from the ability of steel and timber to resist the tension forces which would be developed during a buckling situation. Brickwork has effectively no strength in tension. As it is identical with a brick wall I shall deal first with the isolated brick column or pier.

brickwork

The method to check the load carrying capacity of a brick pier is identical to that of a brick wall. The main point to remember is that, in the case of a rectangular pier rather than a square one, the narrower dimension must be used for effective width as this will be the face which will buckle first.

As an example we will assess the load carrying capacity of a 450 × 225 common brick pier in 1:1:6 mortar, with joists bearing on to its wide face and a height from foundation to joist bearing of 2900mm.

Effective thickness		= 225mm
Effective height		= $\frac{3}{4} \times 2900 = 2175$mm
Slenderness ratio	= 2175/225	= 9·7
Stress Reduction factor (from the table)		= 0·90
Basic Stress (as before) = 1·35 N/mm^2.		
Working stress	= 0·90 × 1·35	= 1·21 N/mm^2.
Design load capacity on pier		= 1·21 × 450 × 225
		= 122 510 N
		= 122 kN
		= 12 200kg Mass

steelwork

Most of the controls and technical works on steel columns apply to the steel stanchion as part of a steel frame and therefore include many more criteria and details than would be applicable to the scale of domestic and commercial work covered by this book. However, individual steel columns are encountered particularly domestically, and you should be able to calculate their carrying capacity if you have to. I have, therefore, geared what follows to these situations.

In general terms beecause of the considerable extra strength of steel compared to timber, brick and concrete, any restraint needs to be both more substantial and is more critical than for the other materials. The main point is that in order to have a FIXED end condition on a column a properly designed steel junction is needed. this will not often be the case here which simplifies matters considerably.

As before the SLENDERNESS RATIO will need to be calculated against which a REDUCED DESIGN STRESS will be read from the table for AXIAL COMPRESSION. Slenderness ratio will be 1/10ry as before. For normal end conditions and with the head completely held to withstand sideways movement in either direction, the effective length, 1, will be the actual length. If the building or that to which it is tied is prefabricated or anything but substantial and rigid, some slight sidesway has to be assumed and the effective length, 1, increases to 1½ times the actual length. The RADIUS of GYRATION is again read from the table for that section of steel and if it is not symmetrical, the smaller figure has to be used. Full axial compression stress is then reduced to the allowable design stress and the result applied to the actual area of steel in the section.

Let us find the load carrying capacity of a 76mm diam. circular hollow section with 3·2mm wall thickness. The column is 2·900m long and is at the corner of a house extension and fully tied to it.

slenderness ratio = $\frac{1}{10ry} = \frac{2900}{10 \times 2.58} = 112$

reduced axial stress (from table) $= 67\ N/mm^2$

area of section (from table) $= 7.33cm^2 = 733cm^2$

∴ allowable load $= 67 \times 733 = 49\ 100\ N$
$= 49kN = 4900$ kg Mass

timber

The calculation for a timber column is as for a steel column but a little simpler.

The situation regarding effective length applies to timber as it does for steel and I suggest you use the same limitations restricting effective length to either actual length or 1½ times actual length. there is a further case with timber that might arise in some situations. If either end of the timber column, in a sidesway situation, is butt ended with no proper fixity, then the effective length should be taken as **twice** the actual length. Radius of gyration for a timber column in anything other than a single rectangular section has to be worked out and this is $\sqrt{(I/A)}$ where I is the moment of inertia and A is the cross-sectional area. This will not often occur and I will not deal with it further here. The same applies to a rectangular section but this can be simplified to 1/4 b where b is the smaller of the two dimensions.

When we have calculated the slenderness ratio 1/ry for the section we don't have to reduce stresses with timber. The Code of Practice 112 requires only that the slenderness ratio does not exceed 180 and in general terms it would be prudent not to get too near that figure either.

Axial stress for a grade 50 timber parallel to the grain should not exceed $3 \cdot 0\ N/mm^2$.

Let us therefore see the load carrying capacity of a 100×100 timber column of grade 50 timber in the situation of the example for the steel column.

Slenderness ratio $= 2900/25$
$= 116$ O.K.

Load carrying capacity $= 3 \cdot 0 \times 100 \times 100$
$= 30\ 000\ N$
$= 30\ kN$
$= 3000$ kg Mass

the foundation

concentric foundations

When a wall or column sits symmetrically on the concrete footing or foundation, the foundation is referred to as concentric and you will be pleased to learn that these concentric foundations are a welcome return to simple arithmetic.

The concrete foundation does two jobs. The first and main job is to spread the load coming down the wall before it enters the ground. In other words if the foundation or footing to a wall is twice the width of the wall then the stress into the ground will be half that in the wall.

The principle in foundation design will be to lower the stress in the ground to come within the LOAD-BEARING CAPACITY of that particular ground stratum. This we shall deal with later.The second job of a foundation is to provide a hard and level bearing from which to build the walls.This may seem obvious but for instance when building on rock (which has similar or greater strength than brickwork) one may not need to reduce the stress by spreading the load but one will still need a footing because of the level base it will provide.

Normal domestic foundations are now adequately covered for sizing in the deemed-to-satisfy parts of the Building Regulations and you will not often need to use calculations for foundations. On the infrequent occasions when this will arise the situation will probably be unusual and therefore not covered by the deemed-to-satisfy tables anyway.

In the situation where you are checking the size of a concentric foundation you will have to work out the load coming down the wall or column involved. This will be done by the method described in Chapter Two for that part of the foundations involved. Having achieved this loading figure at the foundation either as kN/m, for a wall, or kN, for a column, it will need to be translated into a kN/m^2 figure to represent the load per square metre being applied at the underside of the foundation to the ground.

For the wall strip footing, the load per metre will be divided by the footing width (in metres) to give the bearing pressure. For a pad footing to a column or pier the load will be divided by the area of the pad (in square metres).

The bearing capacity of the ground will vary according to its nature and in many circumstances a figure will need to be agreed with the local building inspector.

The following table will give a very rough guide but individual circumstances will vary considerably. With something like foundations, as with all structures, always err on the side of conservatism and safety.

hard rock	2000 kN/m²
hard chalk	600 kN/m²
stiff clay, hard sand	200 kN/m²
firm clay............................	100 kN/m²
soft clay	50 kN/m²

In the normal unknown case with no especially solid ground expected to be encountered, you should assume 50 kN/m² until excavations can adjust the matter either way.

eccentric foundations

Whilst eccentric foundations can be rather odd the meaning of the word here is that the load is not central or concentric on to the footing. This situation with high loads can cause big problems and professional help is sometimes needed. However, I think it is appropriate that the fundamental points be explained as it is not uncommon and with ordinary loads the matter can usually be easily dealt with.

Eccentricity in structural terms means the amount by which the centre of action of the force is away from the centre of the support or foundation. This can be measured in distance terms or, more normally as a proportion of the width of foundation.

As with a beam having an off-centre point load, the support nearer the load will logically be greater than the support at the other side. With foundations, as a load moves away from its centre towards an edge the bearing pressure nearer that edge must **increase** and that on the other edge must **reduce** from the uniformly loaded situation. Figure 33 sets up alternative situations diagrammatically.

Figs. 33 (a) (b) (c) (d)

Uniformly distributed loading with zero eccentricity is shown in Figure 33 (a). The other extreme for the purposes of this book is in Figure 33 (b) with the load centre at the middle third position or an eccentricity of 1/6 away from the centre. In this situation the load is producing a triangular pressure distribution as the load position is at the centre of action of a triangle of pressure as we found with flexural compression stress in the timber beam. The point to realise here is that the pressure at the

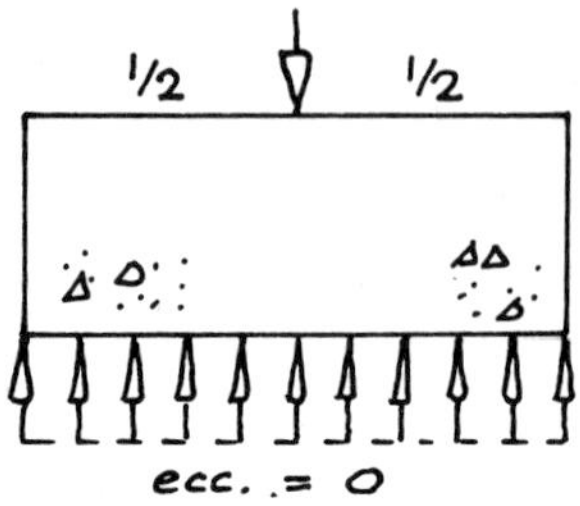

Figure 33 (a)

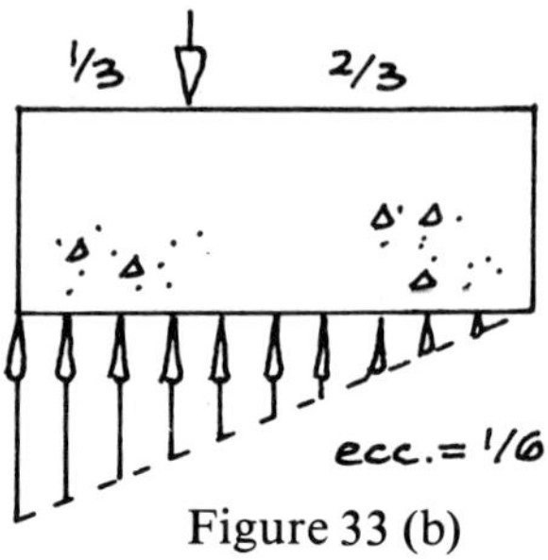

Figure 33 (b)

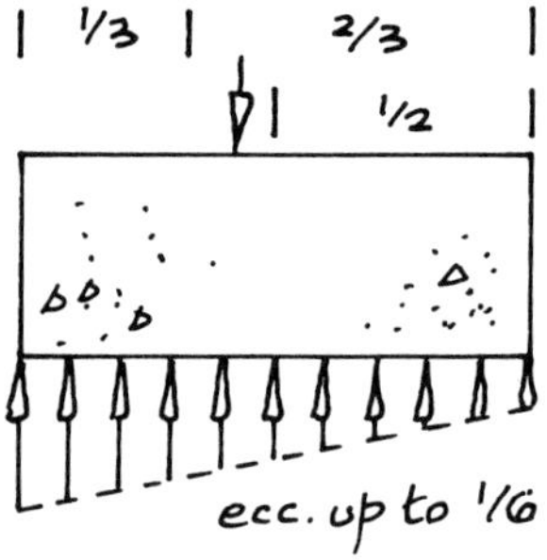

Figure 33 (c)

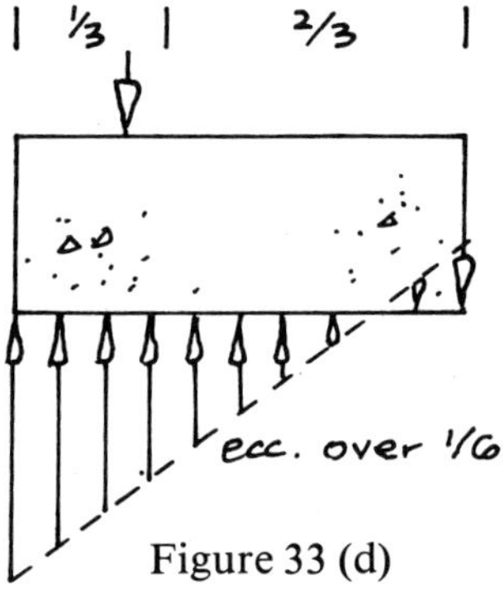

Figure 33 (d)

unloaded edge is zero and that on the loaded edge is therefore double what it was when uniformly distributed. This is because the total load and total thrust are constant and will remain equal. Between these two situations as in figure 33 (c) the pressure variation will be pro rata to the eccentricity between these two extremes.

If the eccentricity increases beyond 1/6, troubles really start as negative pressure should exist as shown in figure 33 (d). This increases still further the bearing pressure at the other side. This is the situation where the structural engineer will consider a ground beam system to balance the eccentricity and I will not deal with it here.

In situations of eccentricity which is often up to a boundary it is worth bearing in mind a deviation from the above. This is that, if the ground bearing capacity will allow, a reduced width of footing will reduce the actual eccentricity to probably a manageable proportion. A development of this is to reduce the footing width to that of the loading wall thickness which will remove all eccentricity. However, this is likely to increase the bearing pressure beyond acceptable limits, but the principle is worth bearing in mind.

Chapter Four

The calculating process

general matters

I refer to the "Calculating Process" quite intentionally to indicate the proper attention needed for this aspect of structures. At this elementary stage and to newcomers it is something which does not often get so treated. As a result when it comes to the fact of a problem in practice, the student or technician, even if he can manage the individual processes already dealt with, cannot put them sensibly together. The word "sensibly" is also important in simple structures and is too often forgotten in amongst the new concepts and language which surrounds the student of simple structures.

There are three quite separate stages in any structural calculation, whether it is a domestic steel joist or a multi-storey frame. These are

1. Assessment and establishing of the loads at the point or on the structural unit in question.
2. Calculation of bending moment caused in that span if applicable.
3. Selection and specification of structural unit to best resist the stresses so caused and at the same time to best fit into the surrounding building construction.

Each of these stages is to be considered carefully and separately as all three play an equal part in the one continuous calculation. Structural engineering even at the elementary level involves the art of practice. Accordingly skill will need to be developed in this art through practice and this will not happen overnight. Initially all that should be sought in a practical calculation is a sound, concise and workman-like exercise. In what follows I have tried to set down some of the more important aspects of the calculating process at this level of construction, points to note and dangers to watch for and avoid.

assessment of load

The first step in considering a structural problem is to carefully assess what parts of the structure are transmitting their loads through to the place in question. This is done by examination and should be a separate and initial matter before calculating the amount of the load. It is very easy, if this is rushed, to forget say a roof load or something similar which may have a material effect on the point concerned. As we have seen, this assessment is done by seeing which way floor joists are spanning, bearing of purlins etc. Having done this then work from the uppermost contributor downwards.

Another factor is non-load bearing partitions. Where lightweight studded partitions occur as they often do in upper storeys, and also solid walls which do not continue downwards, their contribution to the dead load must be included. This is done as normal by starting with a square metre of the partition or wall, and developing it by multiplying by the height to give a load per metre run at the base of the partition. It will depend on the layout of supporting floor joists and the direction in which you are examining load flow as to what contribution the partition is making. With a partition across the joists the load per metre will be transferred to the joist supports in the same way as a point load on a beam, proportionately to the ratio in which it splits the span.

There is no short cut to load assessment and calculation and the arithmetical exercise in bringing load values down the building can only be done carefully and methodically as illustrated. Apart from helping to ensure accuracy it assists both you and others to check what you have included and not included and the details of densities etc., assumed. Similarly any subsequent change in building materials can be easily adjusted in the calculation to check that the previous structural specification remains valid. This part of the calculation will often be, and in some ways, should occupy the larger part of the written work for the calculation and specification of a structural unit.

Another point not to be forgotten and easily is, is the self weight of the beam. Our exercises on bending moment and shear force and their diagrams assumed a weightless beam for clarity. In the real world the weight of the beam has its own contribution to make regarding its internal stresses. Whilst at this stage you don't know the size of the structural unit, you will have an idea at least of its material and a guess must be made to be checked on completion. This guess needs to be high rather than low to save repeating work if the loads you have assumed subsequently prove to be insufficient.

concentrated live load

The Code of Practice CP 3 Chapter V, selections of which comprise Appendix II, makes reference to "Concentrated load" and a word of clarification is needed about this. Live load for a floor is an even load per square metre over the whole area, which allows calculations to be made of general floor strength and the resultant effect through the building. We all know that in fact live load is not even over every square metre of a floor. The way in which loads arise on the floor from actual activity will depend on the use involved. The concentrated load figure in the Code makes recommendations on this matter and the relationship between concentrated load and general live load will vary with the use class of the floor. The concentrated load is shown to be applied over a 300mm square and any checking needed for this is done without addition of the general live load. In normal use, this need not be considered but with anything unusual or heavily loaded, like a garage, or book store, it is used to check the upper supporting skin of the floor only. Examples of this would be the thickness of floor boards if floor joists had to be spaced far apart for some reason or for checking duct or pit covers in a garage.

window openings

The existence of windows in walls is something which at first sight would seem to complicate matters regarding loading calculations. However, in normal use, assume they are not there. The brickwork you will calculate in its place will then be easier and continuous and will also help to incorporate a small extra safety factor to your calculating—never a bad thing if not done to excess. There are exceptions to this. For example if your structural problem is critical as far as size is concerned and you cannot afford the luxury of surplus design load then the saving in weight of brickwork in the window opening will need to be taken. In those situations the actual weight of glass and timber must not be forgotten and must be added back in.

arching in walls

Another major factor regarding loading of and in brickwork is the property of a brick or block wall to ARCH. Arching of brickwork is something you will probably have heard of and is largely self explanatory. All design is based on a failure situation or mode of failure and the appropriate safety factor incorporated into the calculation by way of the DESIGN stresses used. The property of brickwork therefore in ARCHING across an opening without a beam in place at all does bear clarification with regard to the design process. You may imagine that if there is an opening in a brickwall as in Figure 34 and the lintol or beam above it were removed then the brickwork that collapses would not be the full height of brickwork above the opening as shown between the dotted lines but merely that indicated.

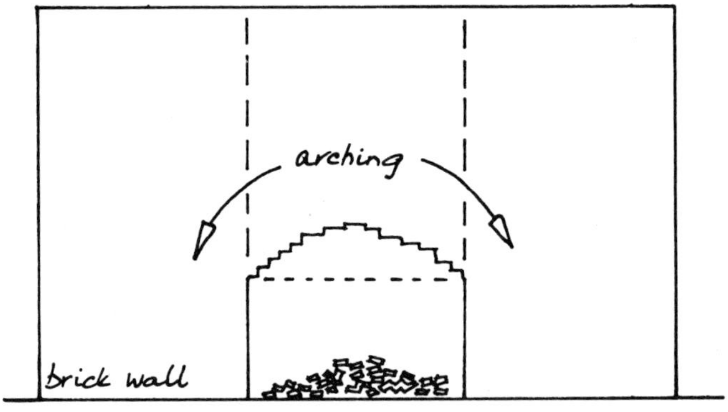

Figure 34

The remainder is supported by the arching of the brickwork transferring the loads above down beside the opening. Whilst we shall see below how to take advantage of this we must not forget one important point. This is, that, in disregarding some loads from above in calculating our beam loads, those loads don't "go away" but are merely "redistributed" by this property of any wall acting solidly together down beside the opening.

You will see that the shape of collapsed brickwork in Figure 34 is

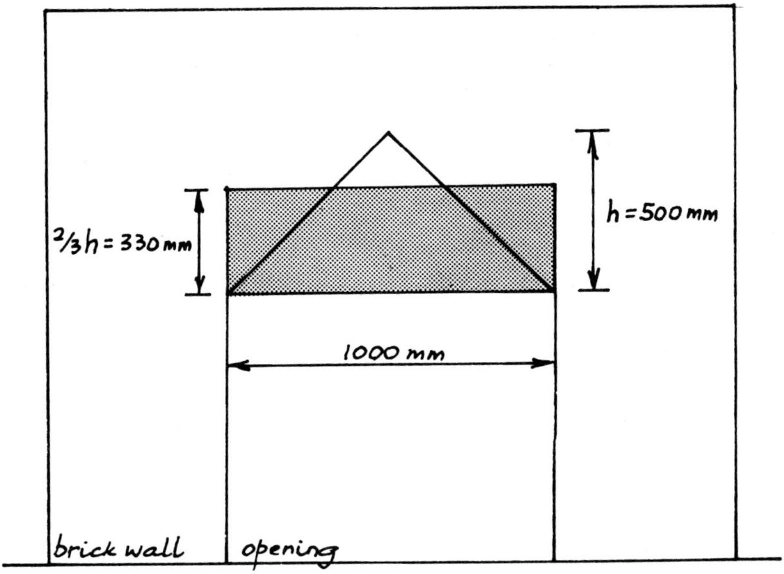

Figure 35

roughly curved or triangular in shape. Whilst a triangular wedge of brickwork can be taken as the load in these situations and the rather complicated bending moment arising calculated, it is usually easier and more efficient to rationalise the failure shape to a rectangular one as in Figure 35.

Use therefore that height of brickwork but rectangular for the full length of the span. This will give you a slightly inflated value for the bending moment which isn't a bad thing. A rectangular area of brickwork ⅔ of that height will give you the same maximum bending moment as the triangular area.

Imposed loads into the wall, like floor joists, within that zone of collapse are also included in the load. The question of imposed loads just above that failure shape is one about which opinion will vary and which experience will affect. The yard stick to use is—if in doubt then include it. However, I hope you will see that there comes a point where even heavy point loads, if they are far enough above the span, are not going to impose loading on to the beam.

There will also be varying opinion as to the degree of arching that can be assumed in any one situation. I will now set down two criteria which I would consider satisfactory in use.

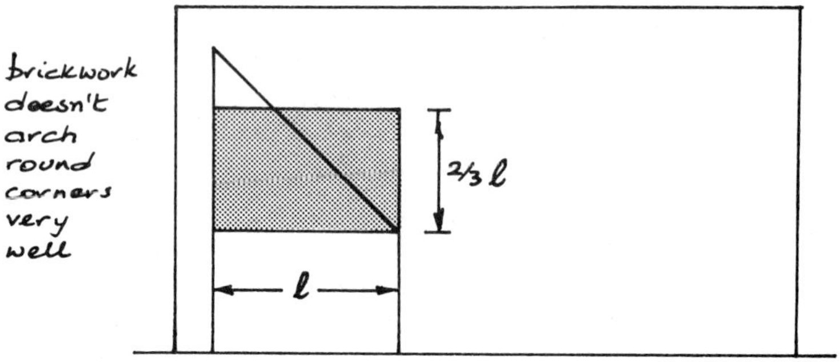
brickwork
doesn't
arch
round
corners
very
well
2/3 l
l

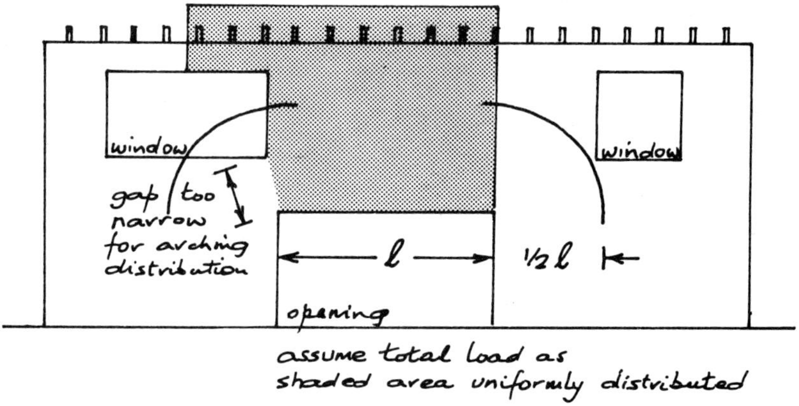
window
window
gap too
narrow
for arching
distribution
l
1/2 l
opening
assume total load as
shaded area uniformly distributed

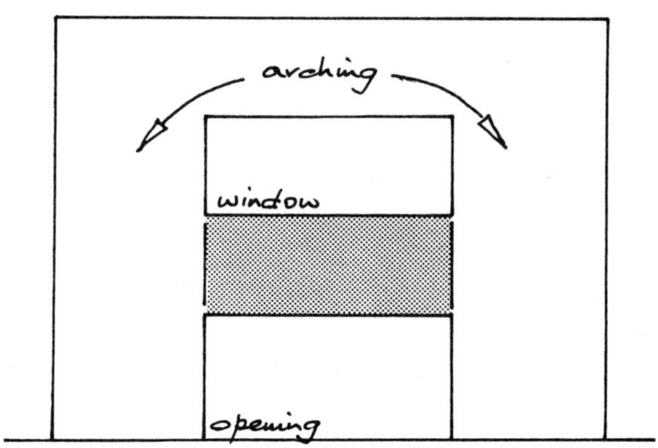
arching
window
opening

Figure 36

1. Brickwork cannot arch very well round corners and therefore assume that it does not.
2. There must be "wall" in the area of the arching process which can distribute these loads. Whilst each case may have peculiarities of its own, a useful guide line to work on is that there must be no interruption to the wall by window or similar discontinuity within a distance from the joist bearing of half the span. This would be above and beside the bearing.

By the same token do not forget loads within a wall from an existing arching situation. The redistributed loads from this may now act to superimpose load on the new opening and will therefore need inclusion. I illustrate in Figure 36 one or two cases to illustrate these points with the shaded areas and superimposed loads within them being those to be catered for by the beam shown.

bending moment and shear force

At the level of general domestic or small commercial work most calculations will involve loading which is uniformly distributed or if not then at least maximum bending moment will be near mid span.

Shear force, at this level of work, will not often enter into structural checks within the structural unit selected. Where it will need checking is in the unusual case where slenderness is such that a big reduction in design bending stress has had to be accepted. This should be unusual as a design bending stress of less than 50% of the maximum ought to be avoided if possible and a bigger beam selected. In the very slender situation the support load or shear force will be high compared to beam depth and a shear stress check will need to be made.

shear stress

Shear stress has not so far been covered but in case the need arises it will now be briefly dealt with.

Shear stress is shear force divided by the area over which it is acting. With timber this is the cross section area of timber being used. With steel it is the area of the web. With reinforced concrete it is width × lever arm. These are now summarised in Figure 37 for the three materials with the allowable shear stresses also shown.

The more common use of the support shear force figure from the calculations is to check that the bearing of the beam can take the stress imposed. As this will often be brickwork, a check will need to be made on the allowable stress from BSCP III (with its 25% allowable increase) and also its contribution to general stress in the wall compared to the allowable stress.

This becomes even more important in a cantilever extension to a beam as the shear force at the cantilever support increases considerably when the length of cantilever extension increases by a small amount.

	shear stress formula	allowable shear stress N/mm^2
timber grade 50	$\frac{F}{bd}$	0·80
steel grade 43	$\frac{F}{Dt}$	108·00
reinforced concrete 21·0 N/mm^2 crushing strength	$\frac{F}{b\,la}$	0·70

Figure 37

deflection

Deflection, like shear stress, is a design characteristic which, whilst not needed in every calculation, can be very important and knowledge of it and how to use it is essential.

When all external characteristics of a beam situation are equal, the remaining variable is the material of construction. In these situations some materials deflect or "sag" more than others. This characteristic of the material is called the MODULUS OF ELASTICITY and is measured in units of NEWTONS/SQUARE MILLIMETRE (N/mm^2). Moduli of Elasticity attributable to our main building materials are as follows;

timber .... 3 000 N/mm^2
concrete .... 9 000 N/mm^2
steel .... 20 000 N/mm^2

Unlike bending moment and shear force, it will serve no useful purpose here to show how the formula for deflection is developed. We are usually only interested in the deflection at midspan, as this is the maximum, and the most common loading situation is uniformly distributed. The deflection in these conditions for a simply supported beam is evaluated by the following formula;

$$\text{Deflection, } \Delta, = \frac{5}{384} \frac{wL^4}{E\,I}$$

In practice you will find that this rather cumbersome formula is fairly easy to remember. In this formula, "E" is the Modulus of Elasticity of the material. "I" is a characteristic of the "shape" of the unit and is called the **MOMENT OF INERTIA** of the section. This is a development from the section modulus "Z". The moment of inertia is the section modulus multiplied by the distance from the neutral axis to the extreme fibres. For a rectangular beam therefore, where Z is $bd^2/6$ (see page 44), the Moment of Inertia, I, is $bd^2/6 \times d/2 = bd^3/12$. As you see this is a function of the shape of the section and nothing to do with what it is made of. Moments of Inertia of steel sections which are of course not rectangular are to be found in the steel tables in the Appendices to this book.

To put this into practice, let us consider the example of a 200 × 100 timber beam simply supported over a span of 3 000mm and carrying a uniformly distributed load including its own weight of 1·2 N/mm.

$$\Delta = \frac{5}{384} \times \frac{1.2 \times 3000^4}{3000} \times \frac{12}{100 \times 200^3} = 6.33 \text{ mm}$$

This is always rather a mouthful to calculate even with a calculator and to reduce error it is better to first simplify and collect the zeros as follows;

$$= \frac{5}{384} \times \frac{1.2 \times 3^4 \times 10^{12}}{3 \times 10^3} \times \frac{12}{10^2 \times 2^3 \times 10^6}$$

$$= \frac{5}{384} \times \frac{1.2 \times 3^4}{3} \times \frac{12}{2^9} \times \frac{10^{12}}{10^{11}} = 6.33$$

design use

Having gone through all this and hopefully obtaining the right answer, what are we doing it for?

The reason is that in building construction anything more than the smallest deflection is a nuisance. This is because most of our building finishes or the "clothing" we put on the structure are brittle and will crack if subjected to unacceptable strains. Deflection when compared to the particular span gives an indication of the relative movement to be experienced. The maximum deflection recommended to be acceptable is $0{\cdot}003 \times$ span.

In our example above we see that the deflection of 6·33mm when related to the span in the same units is $6{\cdot}33/3000 = 0{\cdot}0021$ which is therefore acceptable.

Now, naturally, a deflection calculation and check does not have to be carried out for each beam we become involved with. So when is it done? You will see in the formula for deflection that the length or span is to the fourth power. That means that when the span of a beam is doubled, the midspan deflection is increased sixteen times. Compare this to bending moment involving only the squared power of the span which would only increase four times when the span is doubled. Deflection would only normally be calculated and checked therefore when the span is long for the section being used. In other words, light loading over a long span will only require a fairly small section to withstand the bending moment, but deflection may be such as to cause surrounding problems. In that event the section size used would need to be increased beyond that needed for strength. It would then have a larger Moment of Inertia and the deflection would be reduced.

In an extreme case, it may be necessary to change the material, say from timber to steel, and take advantage of the much greater Modulus of

Elasticity ("E" value). Whilst unusual, this can happen in, say, a long span roof purlin where intermediate support might not be available and a sag in the roof surface would result.

material and unit selection

You will usually have some idea in practice as to the material you will use for a particular situation. You will find that the three basic materials tend to pair into two from this point of view. Timber and steel, being both "dry" in construction, tend to go together, with steel being introduced when the strength of an economically sized timber unit is not sufficient. Keep your mind open in these situations to consider not only a change from timber to steel but also from steel to timber when only small stresses are involved as the cost difference is considerable. Reinforced concrete will usually be a predetermined material by reasons of general building characteristics. This would be perhaps in external walls where a boot lintol is required to neatly support the outer skin. These situations are now being reduced by the use of prefabricated steel lintols which give an even neater job and comply with the thermal requirements of the Building Regulations with less effort.

One of reinforced concrete's benefits not yet mentioned is its built-in resistance to fire and the equivalent fire resistance sections of the Building Regulations. Timber and steel have to be separately insulated against fire where required. Ordinary concrete itself can also be used as an insulation against fire for steelwork. We shall see that in this case the concrete encasing also provides adequate lateral restraint to allow the use of maximum steel stress. The extra weight of the concrete however reduces this advantage somewhat compared to dry lining.

design principles

We have seen that at the point in our calculation of unit selection and specification we are at the stage when we are looking at a maximum bending moment for which we have to cater. Whilst I deal shortly with points to note in this respect for the three materials, a general comment now is appropriate.

Practical structures, or building itself for that matter, is not an exact science and in our structural design we must not act as if it were. Naturally we have to be as exact as possible in the calculation of our loads on to a beam and accurate in the subsequent bending moment and shear force calculations. The selection of our beam size however is an operation that establishes that the beam is **strong enough**. Its moment of resistance will never exactly match the applied bending moment but should be in excess thereby ensuring a further small increase in safety factor. Particularly with the smaller units where width and size may be fixed by building construction requirements, the moment of resistance may be considerably larger than the bending moment. This is perfectly satisfactory. However in larger situations, longer spans, heavier loads, the closer one can match moment of resistance to the bending moment, the more economical is the structure. Here again though, experience, skill and commonsense will have to become your supervisors.

practical considerations

timber as beams

As with most structural design the selection of a suitable size will be one of trial and error. If you have tested a particular size and found its moment of resistance is not sufficient for the bending moment, don't cross it out. Merely note that this is so and proceed with testing another size. By then you will have a fair idea as to how much larger a size you will need and you should not, therefore, need more than two attempts. Bear in mind that, from the formula for moment of resistance, doubling the width will double the moment of resistance but doubling the depth will increase the moment of resistance by four times. In other words for the same area of timber section, the deeper it is compared to its width, the more efficiently it is being used. As building timber is most easily available in 50mm widths it is worth remembering that two 50mm wide beams bolted together are a strength equal to a 100mm wide beam and may be less expensive. The use of pairs of beams is quite usual as we shall also see with steel but it is always wise to bolt them together through their neutral axis at horizontal centres in the region of three times the depth. This ensures that any loads which may come more on one side than the other are supported by the whole structural unit as calculated, rather than by half of it which may not be sufficient.

In assuming a grade 50 timber in your calculations make sure that that grade is also specified on the drawings. This may seem obvious but the Grade provisions carry quality requirements apart from the size, which are assumed by the British Standards Institution in recommending the design stress. For instance, excessive knots and shakes will weaken timber. This may be satisfactory for ordinary use but not for a structural unit.

steelwork as beams

Whilst the steel section which one first thinks of and in fact is most used is the I section or joist, there is another section, the channel, which is also very useful. Channels and their moduli are calculated in exactly the same way as joists and can also be equally paired together as mentioned later.

The process of steel beams selection is one of matching design stress × modulus (which gives moment of resistance) with the applied bending moment. Because of the nature of the material and its manufacture we are in fact looking for a suitable modulus in the tables based on an assumed and fixed design stress. In practice, therefore, we can adjust the formula of $M = fZ$ where M is the moment of resistance to, Z (required) $= M/f$ where M is the bending moment. This will give us a direct connection between bending moment and modulus allowing us to go straight to the tables. This does assume however knowledge of the value of "f" in the particular case. This is affected by adequate lateral restraint. Let us assume for the time being that we do not have adequate lateral restraint. This will be the case more often than not, at this level of construction.

So that we can progress even though we have two unknowns – "Z" and "f" – we have to estimate a value for "f" and proceed to see what "Z" value would be required. From the table we then obtain a steel section whose "Z" value or modulus is nearest to and greater than the

"Z" (required) value. Having then achieved a suggested beam size, we are in a position to look up D/T and 1/10ry for the particular span and from the allowable stress Table in Appendix Vll find the design stress permitted. Comparison with the stress we had initially estimated will show us whether we had guessed too high or too low. If we had guessed too high a bigger section will be needed, and the design stress checked again. If we had guessed a bit too low then everything is satisfactory. If our guess was considerably lower than necessary it may mean that a smaller section will suffice. In that case try something smaller and re-check D/T and 1/10ry.

The use of two joists or two channels side by side often creates a tidy solution to many problems. We often need to support a cavity wall with a beam of only 100mm or 150mm width. The use of a plate on top to increase the width of support is often difficult and untidy. The use of two beams of smaller section bolted together usually provides a tidier and more satisfactory solution. We shall also see that two channels bolted back to back can often provide the best solution in a lightly loaded long span because of the extra stiffness created.

In using two beams bolted together each will be carrying half the moment and each may then half the modulus required. We also have an extra advantage of stiffness. Some people do assume that in bolting two beams together each provides the other with adequate lateral restraint and therefore maximum design stress may be used in both. Whilst this approaches the truth it is not strictly correct and can be dangerous. The true situation which should always be used is that the radius of gyration of the double beam acting together is twice that of the individual value. D/T will be the same and 1/10ry should be calculated using the larger value of ry. You will find that this large reduction in slenderness ratio will usually allow a maximum design stress but in very slender cases this will not quite be the case.

Adequate lateral restraint in practice is something we must look at and opinions may vary as to whether a particular form of construction will be providing adequate lateral restraint to the steel in question. I have set down diagrammatically in Figure 38 some typical forms of construction which in my opinion do or do not provide adequate lateral restraint.

Whilst I hope you will get some idea from these examples the sort of situation that does and does not provide adequate lateral restraint, a word or two of reason may assist.

The most frequent relationship with steelwork in domestic and commercial work is with timber joists. A handy detail is that of wedging the joists into the web to allow a clean ceiling line. As long as the timber joist depth is no less than that of the steel joist it can be assumed as providing adequate lateral restraint. When the depth of the steel joist increases beyond that of restraining timber joists the forces tending to

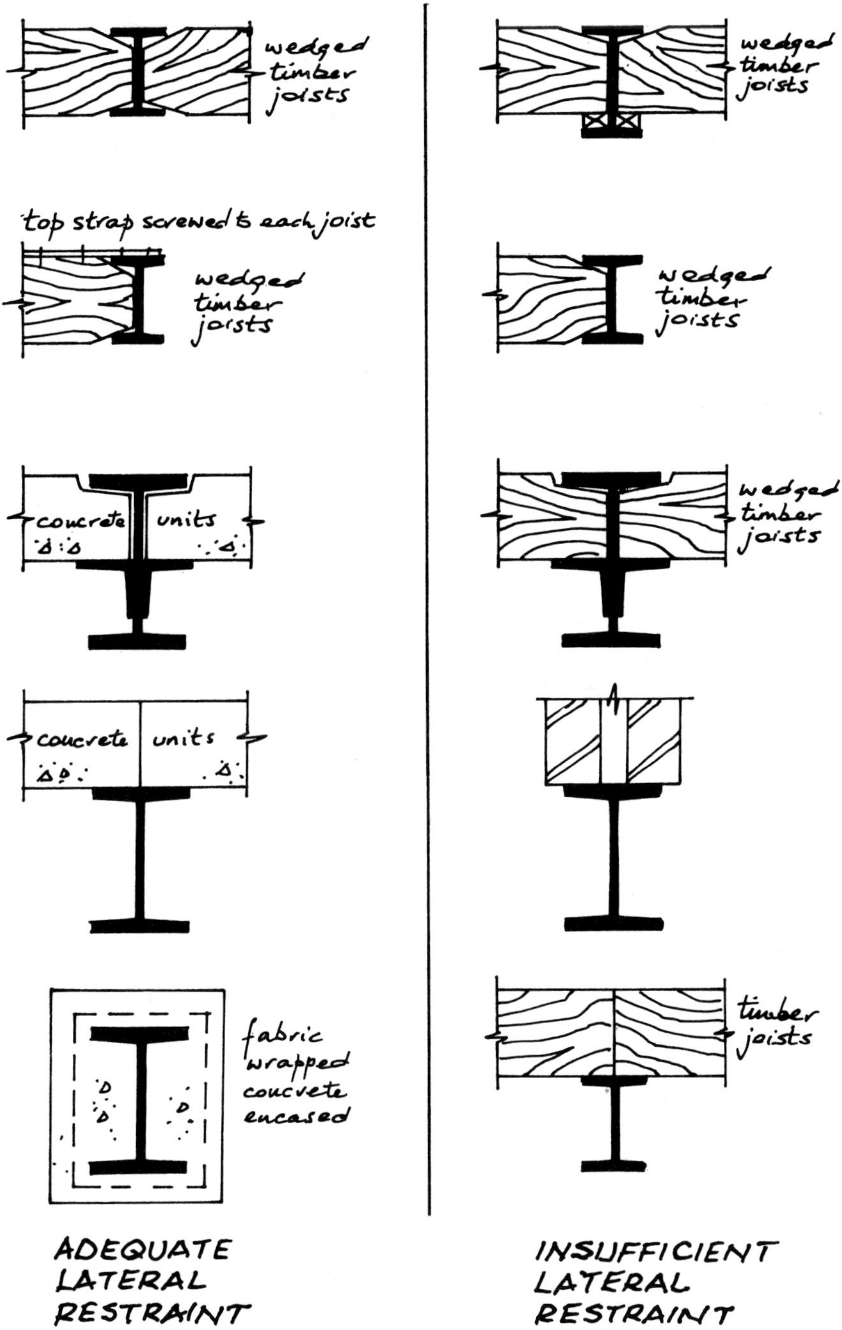
wedged timber joists
top strap screwed to each joist
wedged timber joists
concrete units
concrete units
fabric wrapped concrete encased
ADEQUATE LATERAL RESTRAINT
wedged timber joists
wedged timber joists
wedged timber joists
timber joists
INSUFFICIENT LATERAL RESTRAINT

Figure 38

create a sideways buckling action will be that much greater and at the failure situation can crush the insufficiently strong timber joists. The same construction with precast concrete units will be stronger and generally adequate. Even here though there must be a limit to the depth relationship which would pertain in a very long span steel beam supporting short and shallow concrete units. The situation, however, would be rare in practice.

reinforced concrete as beams

In domestic and commercial work the majority of reinforced concrete beams are in the form of lintols over openings for windows and doors. The majority of these cases are well supplied by the precast concrete industry through the contractor or builder and need not very often involve the specifier in calculations.

The usual case involving calculations would be where reinforced concrete cast in position can do a job which, for one reason or another, steel or timber cannot.

The principles of the calculations to ascertain both concrete strength in bending compression and steel strength in bending tension have been dealt with. An important factor has been assumed in those explanations but has not been referred to. This factor only becomes necesssary in larger sizes of unit. It relates to the way in which the steel and concrete act together to provide a joint resistance to the bending moment applied.

There are two aspects to this. One is that the steel is "gripped" by the surrounding concrete. This is called BOND and the stress created round the surface of the bar is called BOND STRESS. To illustrate this imagine the steel reinforcing bar in a cardboard sleeve separating contact with the concrete. In this situation it would obviously not be "stretched" or stressed by the bending action, nor, therefore would it be contributing to the strength of the beam. The other is the connection between this concrete gripping the steel and the concrete in the top half in compression. Similarly if there were no such connection (as in a cracked situation) no combined strength as calculated would result.

Bond strength in practice is catered for by splitting up the total steel tension area into a number of bars relating to the width of the beam. This increase into a larger number of narrower bars increases the total surface area of steel and therefore usually keeps the bond stress within bounds. Similarly maximum bond stress will be at maximum bending moment. The steel in the ends of the beam with low or zero bending moment will therefore be quite happily "anchored" there and so allowing steel stress to build up where it is needed in mid span. This anchoring would not be so effective in a short span beam and in these cases it is often wise to bend up the ends of the bar in a "hook" which properly anchors the bar in a short length.

Connection between the top and bottom halves of the beam, as concrete is weak in tension, has to be reinforced in larger beams by "links" or "stirrups". These are narrower diameter steel rods bent into a square and positioned across the line of the beam at centres roughly equal to the depth of the beam. The tension steel is fitted into the bottom corners of these links and wired to them prior to concreting. To ensure stability of the "cage" created, small diameter bars are usually also wired into the top corners. In the wider reinforced concrete beams, which are not unusual, you may have three or four or more bars along the bottom of the beam. The Code of Practice does require that each reinforcing bar has a link in contact with each at two points at right angles. With two bars, one in each corner, this is clearly dealt with by "single" links at the appropriate centres. Three bars though will require double links at each position each just over half the width of the beam, one to one side and one to the other. This is illustrated in the worked examples later on.

Chapter Five

Worked examples

All these examples of working procedures are taken from examples I have come across and are, therefore, you may think surprisingly, real practice situations. I have referred to them as "Tasks" which further demonstrates their realism.

Task One

This client had been a bit too clever for his own good and had done his own structural alterations. The Local Authority had asked for calculations to be done to prove the effectiveness of his solution which as you will see was rather short of the mark.

I hope that you will be able to follow the calculations shown but to assist and to make pertinent points of clarifications or comment, I have numbered each line of the working and now refer to them where appropriate.

line 1	Job title and date for each calculation
line 3	Identify the particular calculation.
line 5	Notice that the "loading" list always starts at the uppermost item that is transmitting load on to the position in question.
line 7	Load per unit AREA
line 8	Load per unit LENGTH along the direction needed. In this case by multiplying the two half spans contributing load to the position. Note that the description is also indicated.
line 11	Load per unit area of partition.
line 12	This translates to load per unit length (of the partition) by multiplying by the height of the partition.
line 16	Unit load of the floor starting at the top floor boards and ending at the bottom with the ceiling underneath.

JOB TITLE: Task One DATE

To check 150×150 baulk timber across ground floor

Loading: ceiling: joists 100×50 = 7 kg/m²
pbd & skim = 22
29 kg/m²
load on to partition = 29 × 6000/2 = 87 kg/m
partition: studding 75×50 = 5 kg/m²
2× pbd + skim = 44
49 kg/m²
load on to beam = 49 × 2.300 = 113 kg/m
floor: boarding = 10 kg/m²
joists 200×50 = 14
pbd + skim = 22
46
live load = 153
199 kg/m²
load on to beam = 199 × 6.000/2 = 597 kg/m
Total load on to beam = 797 kg/m
= 7.97 N/mm

Span 2700 mm

$$b.m. = \frac{wl^2}{8} = \frac{7.97 \times 2700^2}{8} = 7\,262\,662 \text{ Nmm}$$

m.r. 150×150 timber = 2 812 500 Nmm – insufficient

Suggest R.S.Channel each side of beam bolted through. Assume for simplicity that the R.S.C.'s will take all of the load and as they will be bolted through at regular centres, it will be assumed that there is adequate lateral restraint and full flexural stress can be developed.

$$Z\ (\text{reqd.}) = \frac{7\,262\,662}{165} = 44016 \text{ mm}^3 = 44.016 \text{ cm}^3$$

2 no. 152×76 R.S.C.'s provide 2× 22.77 cm³ = 45.54 cm³
OK.

10 mm bolts @ 800mm crs bolted through

line 17 It is a convention to keep the live load separate in this way.

line 19 Extension into load per unit length by multiplying by total double span divided by two as line 8.

line 20 The collective total of the separate loads per metre. Note the importance of setting out these subtotals at the far right hand side and then the change to FORCE per unit length (N/mm).

line 22 It is always wise to state the span assumed so that it can be checked against the drawing.

line 23 The applied bending moment calculation. Don't forget the units on the end. Note how words play an equally important part as figures in good calculations. Remember that calculations should be able to be read as **English language**. In other words, because of checking and back reference, they are not just a means to an end but have a permanent value as do the detailed drawings.

line 25 Do not do calculations in rough first. Where your first attempt at the size of a supporting member has proved inadequate, don't be frightened to say so and carry on to some other solution. Doing it in rough first only wastes time and allows mistakes in copying.

line 27 Carry the reader with you by saying what you are doing. This particular solution has obvious constructional advantages compared to replacement.

line 28 This sort of assumption saves a lot of design time and rarely results in a larger beam than would have been the case from a combined calculation. Dangers in faulty design assumptions can also occur with existing materials and unless absolutely necessary are best avoided as here.

line 30 This is also where asssumptions need to be made regarding adequate lateral restraint so say so and in this case "why".

line 33 Remember $M = fZ$, therefore $Z = M/f$. Remember to note the units of mm^3 and then to adjust to cm^3 to coincide with the steel tables.

line 36 The final solution of successful design size can be underlined on completion to make it stand out for quick future reference.

line 38 There are often design specification notes to record at this stage arising out of the assumptions made earlier (line 29). This properly links the design exercise with the following architectural detailing and specification. If not followed up properly in this way they can be left out in error and the assumptions made become unsupported on site.

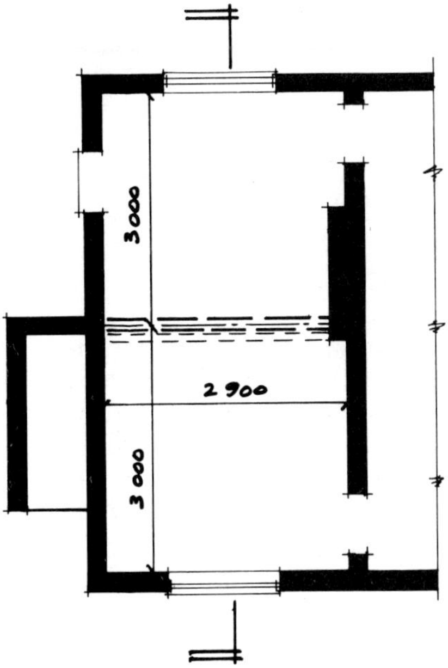

GROUND FLOOR

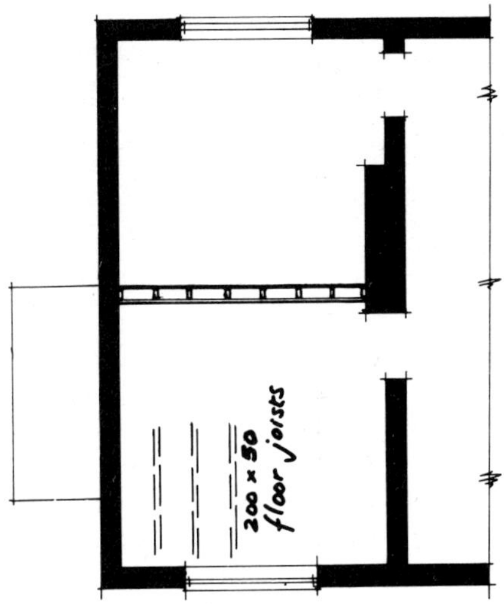

FIRST FLOOR

1

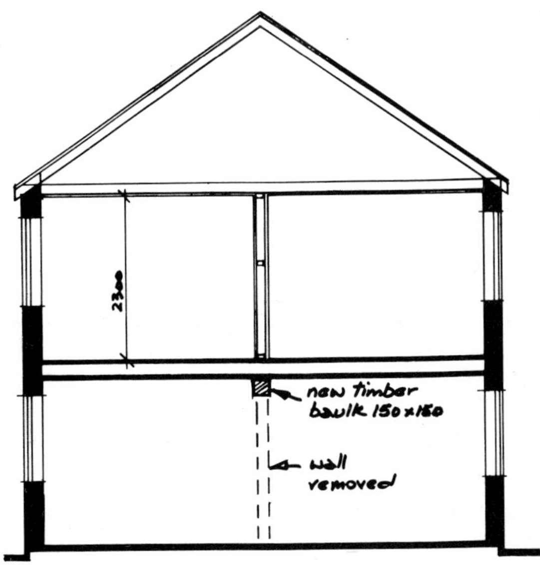

SECTION

Task Two

This was an extension of a lounge in a dormer bungalow and is a demonstration of a lightly loaded long span.

line 1 Job title and date.

line 3 Describe the particular calculation.

line 6 Loading is always the first word in a calculation followed, in this case, by the subdivisions for each area of the construction which is transferring load on to the line in question. Notice here that by examination of directions of span of joists and rafter very little load is involved in the line to be calculated.

line 7 It is useful to restate sizes as a quick check in case something is changed later on.

line 10 Translation of load per unit area into load per unit length by multiplication by the height of the wall in question. Final units to be stated at the end.

line 12 Calculation of unit area weights as with the overall loading always going top to bottom e.g. starting with the felt and chippings and finishing up with the ceiling underneath.

line 17 It is a convention to keep live load separate in this way.

line 19 Extension again into linear load by multiplying the load per unit area by half the span. Always say what the particular calculation is, as shown. Remember that calculations, even though they have a lot of numbers involved, should always read as **English language.**

line 21 Providing the setting out of the calculations has been done properly, the cumulative figure can now be easily added up from the figures on the right hand side.

line 22 Immediate translation into Force per unit length (N/mm).

line 23 It is always useful to state the span assumed for checking purposes.

line 26 This situation is not unusual in domestic situations and so assumptions have to be made and stated. Do not do these trial attempts in rough first. The eventual rewriting wastes time and risks mistakes. Let the whole calculating process of trial and error be seen and hence understood.

line 29 This is an assumption guessed out of the tables as a first try.

line 30 $M = fZ$. Therefore $Z = M/f$. Remember to state units and then adjust to suit the steel tables (cm^3).

JOB TITLE Task Two DATE

To calculate beam as replacement of original external wall to allow lounge extension

Loading: stud wall: timber cladding = 9 kg/m²

studding 75×50 = 5

pbd + skim = 22

36 kg/m²

load on to beam = 36 × 2.900 = 104 kg/m

load from extension roof

3 layer felt = 24 kg/m²

chipboard = 15

joists 100×50 = 7

pbd + skim = 22

68

snow load etc = 70

138 kg/m²

load on to beam = 138 × 1.900/2 = 131 kg/m

self weight of beam + casing, say. = 30 kg/m

Total load on to beam = 265 kg/m

= 2.65 N/mm

Span 5900 mm

$$bm = \frac{wl^2}{8} = \frac{2.65 \times 5900^2}{8} = 11\ 530\ 813 \text{ Nmm}$$

With this long span and low loading and no adequate lateral restraint, try double channel which gives greater total width for same depth.

Assume allowable stress reduction anyway to 130 N/mm²

$$\text{Total } Z \text{ (reqd.)} = \frac{11\ 530\ 813}{130} = 88\ 699 \text{ mm}^3 = 88.70 \text{ cm}^3$$

Two channels 254 × 89 give combined figure of

$Z = 2 \times 45.52 = 91.04 \text{ cm}^3$ OK

Check stress: $\frac{D}{T} = 18.7$, $\frac{l}{10 r_y} = \frac{5900}{2 \times 10 \times 2.58} = 114$

From Appendix VII (Table 3a, BS449) $f = 158 \text{ N/mm}^2$

Because this is noticeably higher than initial assumption (130 N/mm²) try again at $f = 155 \text{ N/mm}^2$

$$\text{Total } Z \text{ (reqd.)} = \frac{11\ 530\ 813}{155} = 74\ 392 \text{ mm}^3 = 74.39 \text{ cm}^3$$

2 × 203 × 89 RSC's give $Z = 2 \times 37.94 = 75.88 \text{ cm}^3$

OK.

line 35 With this method always check the actual allowable stress from BS449. This can be too low or too high.

line 38 This part is not essential as the first try is strong enough but as you see something smaller is quite adequate.

line 40 $Z = M/f$ again with a revised value for 'f'.

line 46 This time the allowable stress of 162 N/mm^2 is still higher than that assumed of 155 N/mm^2 but a better balance has been achieved.

line 48 These long span lightly loaded situations can cause problems at the supports and in deflection. In this case, a good idea to check.

line 49 Total load on the bearing.

line 51 A concrete padstone will be necessary to spread the concentrated load.

line 52 Load divided by area gives stress – in this case bearing stress.

line 54 Now to check the allowable stress in this case.

line 55 Three quarters the height can be used as it is held top and bottom.

line 56 $\frac{7}{8}$ the thicknesses of each part of the cavity wall.

line 60 Basic stress comes from the materials being used.

line 61 Allowable stress uses Basic Stress multiplied by stress reduction factor from table using the slenderness ratio.

line 65 This is the formula for mid span deflection under uniformly distributed load and is represented by the Greek letter Δ. "E" for steel is 200 000 N/mm^2 and the moment of inertia of the section is from the tables. Remember to keep the units the same in the whole expression (mm). Moment of inertia from the tables is in units of cm^4 which has to be multiplied by 10 000 to convert to mm^4.

Check stress: $\frac{D}{T} = 15.8$, $\frac{l}{2 \times 10 r_y} = \frac{5900}{2 \times 10 \times 2.64} = 111.74$

From Appendix VII $f = 162$ N/mm^2 OK

Check bearings

Total load on bearings $= \frac{wl}{2} = \frac{2.65 \times 5900}{2} = 7817$ N

Assume concrete padstone 300 mm long (× 100 wide)

Stress under padstone $= \frac{7817}{300 \times 100} = 0.26$ N/mm^2

Check allowable stress

Effective height = 0.75 × 3000 = 2250 mm

Effective width = 0.67 × 220 = 147 mm

∴ Slenderness ratio $= \frac{2250}{147} = 15.3$

Stress reduction factor (Appendix XV) = 0.75

Basic stress (Appendix XIV) = 1.30 MN/m^2 (same as N/mm^2)

∴ Allowable stress = 1.30 × 0.75 = 0.97 N/mm^2 OK

Check deflection

Deflection $\Delta = \frac{5}{384}\frac{wl^4}{EI} = \frac{5}{384}\frac{2.7 \times 5900^4}{200\,000 \times 2 \times 2491 \times 10^4}$

$= 4.27$ mm

Ratio to span $= \frac{4.27}{5900} = 0.0007$

and is well within the limit of 0.003

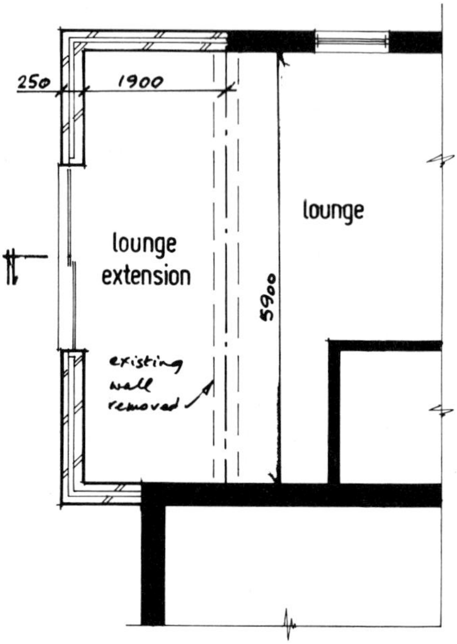

GROUND FLOOR PLAN

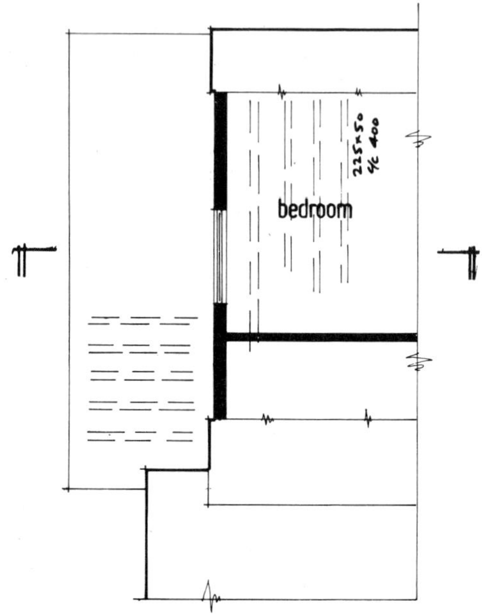

FIRST FLOOR PLAN

2

SIDE ELEVATION

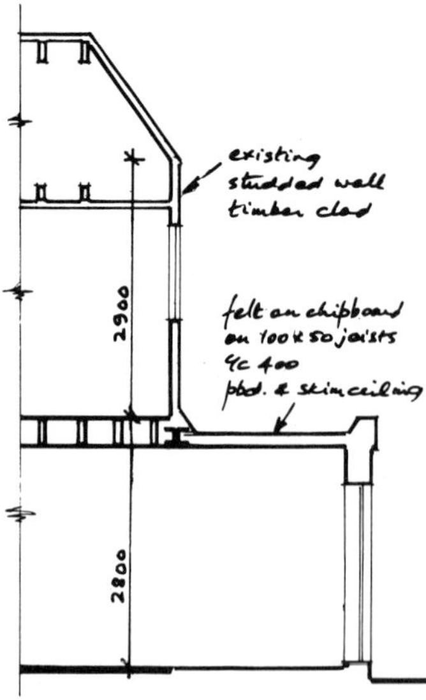

SECTION

Task Three

Alteration of premises for formation of a bank requiring offices above on the first floor.

line 5 No chance of load shedding by arching here. All the roof load will come down through the front wall on to the beam.

line 11 Don't forget to measure the roof surface up the slope.

line 12 For the same reason it is best to take the ceiling separately from a sloping roof as this will be measured horizontally.

line 20 Measure the height of the actual wall construction which will be well up into the eaves and down on to the beam itself. As you can see this wall construction is the major part of the load on to the beam.

line 28 Large beams and particularly when concrete encased will have a self weight which is a material part of the total load as here and so must not be forgotten.

line 29 Eventual size must be guessed (oversize) and then check back on completion.

line 32 Technically the span of a beam should extend past the edge of the support on each side by the distance of the lever arm. I have only gone some way towards that here because the situation will be offset by the extra stiffness at the supports due to the building in of the beam. It is always best in any beam calculation to work on a span slightly larger than the clear dimension as described above.

line35 The validity of this needs to be consolidated by wrapping the steel in a special light mesh prior to concreting to give the tension resistance to the concrete to resist the compression buckling of the steel as now assumed.

line 44 If the calculation is ahead of architectural detail your overall knowledge must be used in this way so that your assumptions and calculations have a chance of remaining valid through to the construction stage.

line 50 The buckling weakness of the brick pier is perpendicular to the side walls and so that is the situation to check. If the pier extended out some way from the side wall with the beam on the end of it then the 330 thickness would dominate as the weakest and therefore the design situation.

line 60 I have adjusted the ground conditions here to be favourable to us. This is a situation which otherwise may need a structural engineer, as more complicated ground support construction m be required.

JOB TITLE Task Three DATE

To calculate front beam concrete encased

Loading: roof: clay tiles = 70 kg/m²
batten + felt = 5
rafters 100 × 50 = 7
82
snow load = 70
152 kg/m²
load onto front wall = 152 × 2.500 = 380 kg/m
ceiling: rafters 100 × 50 = 7 kg/m²
pbd + skim = 22
29 kg/m²
load onto front wall = 29 × 3.500 = 101
front wall: brickwork = 225 kg/m²
blockwork = 115
plaster = 22
362 kg/m²
load onto front beam = 362 × 3.000 = 1086
floor: boarding = 10 kg/m²
joists 225 × 50 = 15
pbd + skim = 22
47
live load = 255
302 kg/m²
load onto front wall 302 × 3.500/2 = 528
self weight concrete encasing
say, 0.45 × 0.30 × 2400 = 324
Total load onto beam = 2419 kg/m
= 24.19 N/mm²

Span 6000 mm

$$\text{b.m.} = \frac{24.19 \times 6000^2}{8} = 108\,855\,000 \text{ Nmm}$$

With concrete encasing assume full steel stress. $f = 165$ N/mm²

$$Z \text{ (reqd.)} = \frac{108\,855\,000}{165} = 660\,000 \text{ mm}^3 = 660 \text{ cm}^3$$

UB. 356 × 171 × 45 kg/m gives 686.9 cm³ OK

Check supports

$$\text{Support load} = 24.19 \times \frac{5700}{2} = 68\,941 \text{ N}$$

Concrete padstone over whole area except front skin of brickwork

Area = 450 × 210 = 94 500 mm^2

Actual stress = $\dfrac{68\,941}{94\,500}$ = 0.73 N/mm^2

Check allowable stress

Brickwork column 450 wide

Effective height = 0.75 × 2500 = 1875 mm

Effective thickness = 450 mm

∴ Slenderness ratio = $\dfrac{1875}{450}$ = 4.2

∴ Stress reduction factor = 1.00

Basic stress (common bwk) = 1.3 N/mm^2

∴ Allowable stress = 1.3 × 1.00 = 1.3 N/mm^2 OK.

Foundation

Assume foundation maintained at 450 wide to avoid eccentricity and permission has been obtained to extend forward 400 mm. Therefore with 600 thick foundation, the spread front to back will be assumed at 1100 mm

ie. area of 0.45 × 1.10 = 0.495 m^2

stress into ground = $\dfrac{68\,941}{0.495}$ = 139.3 kN/m^2

Stiff clay located on examination and approved bearing capacity of 200 kN/m^2 so. O.K. Further distribution of ground stress back along side strip footing will further reduce ground stress.

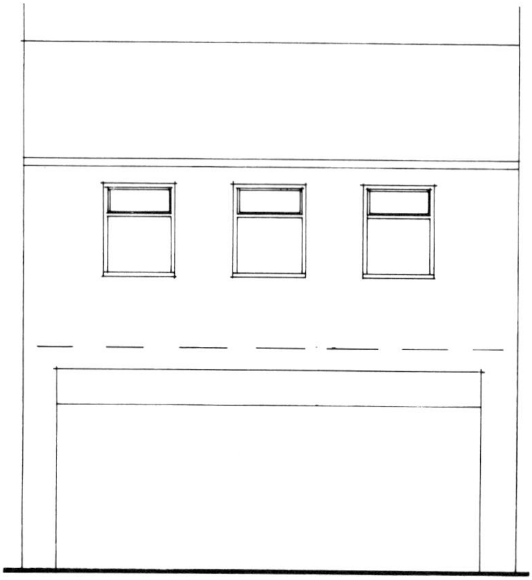

ELEVATION

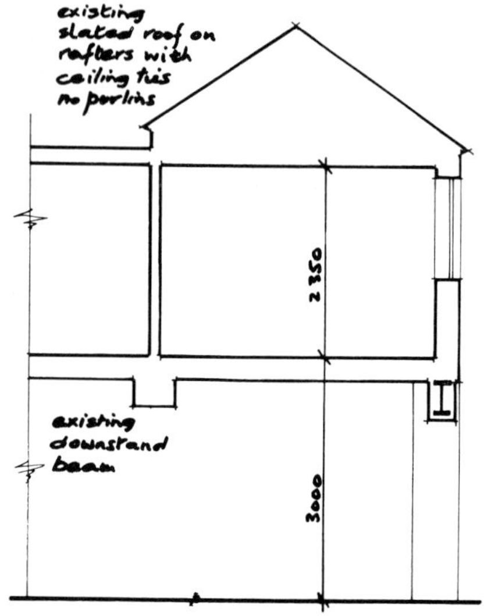

SECTION

3

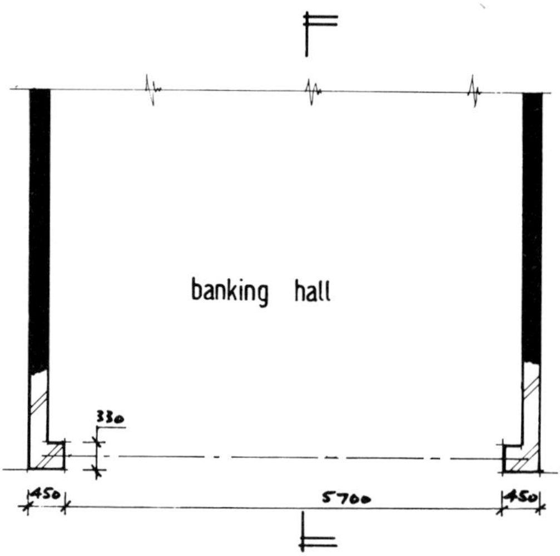

GROUND FLOOR PLAN

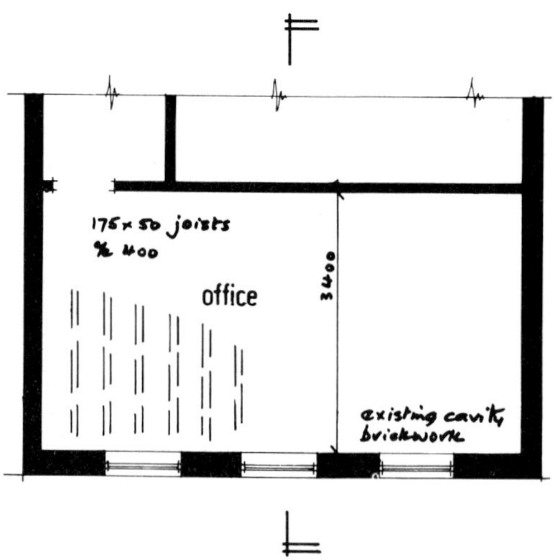

FIRST FLOOR PLAN

Task Four

A rather lengthy structural task of home extension but which also demonstrates that properly set out calculations can bring order and therefore efficiency of time and effort to what otherwise would have been a much more involved operation.

line 3 Without the benefit of a special drawing as here, it is best to draw a small sketch plan of the structural layout, not to scale, so that you can letter up the beams to be calculated. I have started with Beam A after careful examination on the "top-to-bottom" principle and found that Beam A is the uppermost structural element. This then establishes support loads on to other elements of structure in a natural sequence.

line 5 By examination we see that only the side wall and half the floor span is carried by Beam A. The roof, ceiling and front and back walls are carried by Beams B and D. Because the full width of the wall is being supported no arching can be assumed to reduce our load.

line 15 It is always best to enumerate the span divided by two as here rather than writing down just the value of half the span. It makes checking by yourself and others much easier.

line 23 The double beam solution is particularly useful as it is saving headroom from the added lateral restraint and providing proper width support for the double thickness wall.

line 27 Whilst this is still oversize the beams are already small and no significant saving is achieved by further accuracy. Width is needed anyway to support the brickwork.

line 29 On the principle of immediately checking support loads and conditions of any beam, now is the place to set down tidily all the support loads to be used later on.

line 31 Beam C is taking load from both sides BC and CD.

line 48 No need to detail out the unit weight of the wall again. This is our first sign of saving work by proper setting out.

line 52 Anything non-standard by way of loading is always useful to be drawn in this diagrammatic way with loadings and distances as appropriate.

line 53 A bit of experience coming in here sacrificing some academic accuracy for no lack of safety. We have seen that as more loads are applied to a beam the eventual bending moment diagram

JOB TITLE Task Four DATE

Beam A

Loading: wall: 2 skins brickwork = 450 kg/m²
plaster = 22
472 kg/m²
load on to beam = 472 × 4.000 = 1888 kg/m
first floor: boarding = 15 kg/m²
joists 175×50 = 11
asbestos sheeting = 20
46
live load = 153
199 kg/m²
load on to beam = 199 × 1.900/2 = 189
self weight beams say 2×15 = 30
self weight encasing say = 10
Total load on to beam = 2117 kg/m
= 21.17 N/mm

Span 2700 mm

$$b.m. = \frac{21.17 \times 2700^2}{8} = 19\,291\,163 \text{ Nmm}$$

Assume double beam support bolted together with spacer steels to achieve full steel stress, $f = 165 \text{ N/mm}^2$

$$Z\ (\text{reqd.}) = \frac{19\,291\,163}{165} = 116\,916 \text{ mm}^3 = 116.9 \text{ cm}^3$$

Double 127×76 RSJ's give $Z = 2 \times 74.94 = 149.88 \text{ cm}^3$ OK

Support loads on to other beams

On to beam B = 21.17 × 2300/2 = 24 340 N
On to beam C = 21.17 × 5000/2 = 52 920 N
On to beam D = 21.17 × 2700/2 = 28 580 N

Beam B

Loading: roof: slates = 50 kg/m²
battens + felt = 5
rafters 125×50 = 8
63
snow = 70
143 kg/m²
load at eaves = 143 × 3.000 = 429 kg/m

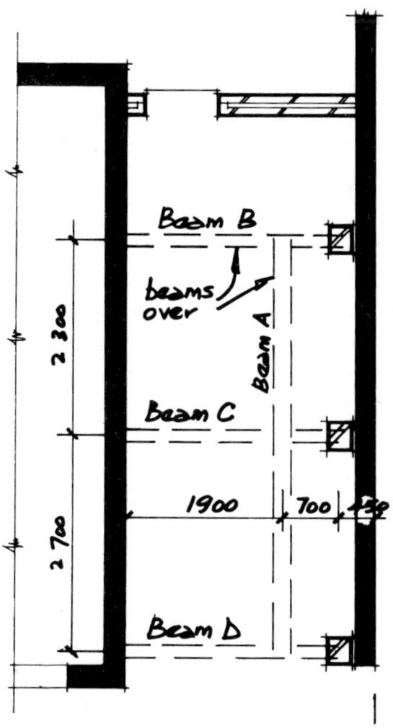

GROUND FLOOR

4

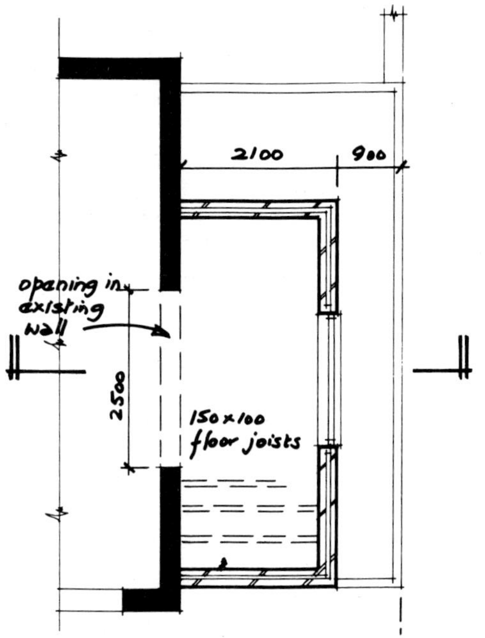

FIRST FLOOR

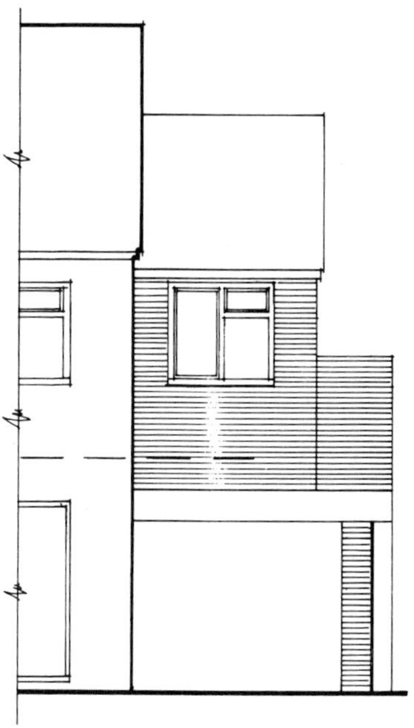

ELEVATION

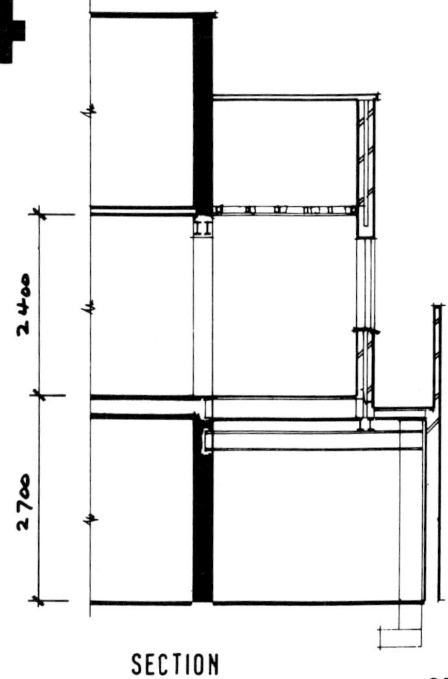

SECTION

approaches the parabolic shape from the single triangular shape. So in practical terms as this point load is near a support the **total** load will be considered uniformly distributed. The eventual margin of size over design modulus will deal quite happily with any small discrepancy.

If the point load position was nearer midspan, then the point load and the u.d. load would have been taken separately. In that situation the u.d. load would have been considered as a point total load acting at its mid-point for purposes of bending moment calculations. Your accumulated knowledge of bending moments produced by different loads will soon allow such safe simplifications to be made. Do **remember** however that the **total** u.d. load needs to be used. It is easy to forget to multiply the length over which it is acting.

line 57 This has not been done before in this book i.e. the use of total load uniformly distributed rather than load per millimetre. All we are doing though is replacing W for wl so that wl^2 becomes Wl.

line 76 Beam D has identical load to beam B and so no further calculation of beam size is needed.

line 79 It will only be necessary here to check the worst load Beam C as the other piers will be the same size anyway.

line 95 We are comparing this 1.27 N/mm^2 with the 0·35 N/mm^2 applied stress in line 83.

line 112 Always **try** to avoid eccentricity. You will find that ground stresses increase dramatically with a small eccentricity. Oddly enough making a foundation smaller to achieve this can often produce a lower stress evenly distributed rather than a high local eccentric stress.

line 119 I have tried to avoid making it easy all the time. Here is another example of not scrapping the whole calculation because the first answer was unsatisfactory.

line 125 Final clarification of size to be used.

ceiling: rafters 100×50 = 7 kg/m²
pbd + skim = 22
29 kg/m²

load at eaves = 29 × 5000/2 = 73 kg/m

wall: as gable wall = 472 kg/m²

load at first floor = 472 × 3.000 = 1416

Total load at first floor = 1918 kg/m
= 19.18 N/mm

Total load = 19.18 × 1900 = 36442 N
point load = 24340
∴ Total load to be assumed u.d. = 60782 N

Span 2600

$bm = \frac{60782 \times 2600}{8} = 19\,754\,150$ Nmm

with stiffening effect of beam A bolted on assume full steel stress, $f = 165$ N/mm²

$Z\,(\text{reqd.}) = \frac{19\,754\,150}{165} = 119\,722$ mm³
= 119.7 cm³

178 × 102 RSJ gives Z = 170.9 cm³ OK

Beam C

Loading: Point load only = 52920 N

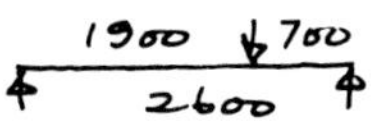

$bm = \frac{Wab}{L} = \frac{52920 \times 1900 \times 700}{2600}$

= 27 071 000 Nmm

$Z\,(\text{reqd.}) = \frac{27\,071\,000}{165} = 164\,067$ mm³
= 164 cm³

as before 178 × 102 RSJ gives Z = 170.9 cm³ OK.

Beam D

As Beam B. 178 × 102 RSJ

Check worst load on pier

Support load Beam C = $52\,920 \times \frac{1900}{2600} = 38\,672$ N

Actual stress on 330 × 330 = $\frac{38\,672}{330 \times 330} = 0.35$ N/mm²

Brick column:

Effective height = 0·75 × 3000 = 2250 mm

Effective thickness = 330 mm

∴ Slenderness ratio = $\frac{2250}{330}$ = 6·8

∴ Stress reduction factor = 0·98

Basic stress for brickwork in 1:1:6 mortar = 1·3 N/mm²

∴ Allowable stress = 1·3 × 0·98 = 1·27 N/mm² OK.

Whilst the stress into the piers is low, it will be prudent to check the other end supports into the existing brick wall:

Worst load, beam C = 38672 N at pier

∴ Load at wall = 52920 − 38672 = 14248 N

Stress under 450 × 100 × 225 deep padstone = $\frac{14\,248}{450 \times 100}$ = 0·31 N/mm²

Slenderness ratio = $\frac{0·75 \times 3000}{2/3 \times 200}$ = 16·8

Stress reduction factor = 0·71

Basic stress = 1·3 N/mm². Allowable stress

Allowable stress = 1·3 × 0·71 = 0·92 N/mm² OK

Foundation

To avoid eccentricity, elongate foundation along boundary wall maintaining 330 width to a length of 600 mm at 300 mm thick

Stress into ground = $\frac{38·672}{0·33 \times 0·60}$ = 195 kN/m²

Ground is only firm clay with bearing capacity of 100 kN/m² so foundation to be increased to 1200 mm long and increased to 600 mm thick to maintain load spread through concrete

Stress into ground = $\frac{38·672}{0·33 \times 1·2}$ = 97·66 kN/m² OK.

Pad foundations to be 1200 × 330 × 600 thick

epilogue

If you have come this far through the book without bloodshed and wonder what all the fuss was about, then I have succeeded and trust that the collective Tables in the Appendices will also allow more convenient practice calculations.

At worst I hope to have exploded what might have been a mystery for you and have shone a light which will grow in strength following your practice in examples.

APPENDIX I

Masses of common building materials by area

Roof tiles laid	clay, plain	70 kg/m²
(area measured parallel	concrete interlocking	50 kg/m²
to plane of tiles)	slates (thin)	50 kg/m²
	cedar shingles	8 kg/m²
	asbestos slates	20 kg/m²
Timber roof battens	25 × 18 (100mm guage)	5 kg/m²
(incl. lining felt)	37 × 25 (100mm guage)	5 kg/m²
Three layer built up felt roofing and limestone chippings		24 kg/m²
Asphalte 18mm two coat		42 kg/m²
Concrete or 1:3 screed	50 thick	120 kg/m²
(2400 kg/m³)	100 thick	240 kg/m²
	200 thick	480 kg/m²
Floor boarding	18 mm softwood boarding	10 kg/m²
	18 mm chipboard sheeting	15 kg/m²
Joists or rafters softwood c/c 400 mm	100 × 50	7 kg/m²
(area measured parallel to	125 × 50	8 kg/m²
plane of the joists)	150 × 50	10 kg/m²
(for joists or rafters at 600 mm crs	175 × 50	11 kg/m²
take 2/3 of these figures)	200 × 50	14 kg/m²
(for joist thickness 37 mm	225 × 50	15 kg/m²
take 3/4 of these figures)	250 × 50	16 kg/m²
	275 × 50	18 kg/m²
	300 × 50	21 kg/m²
Brickwork, clay half brick thick	(one skin 100mm)	225 kg/m²
Brickwork, clay one brick thick	(two skins 200mm)	450 kg/m²
Stone, portland,	100mm thick	224 kg/m²
granite,	100mm thick	264 kg/m²
Blockwork, lightweight concrete		
to BS 2028, Type A	cellular 100 thick	115 kg/m²
	150 thick	175 kg/m²
	200 thick	230 kg/m²
	solid 100 thick	130 kg/m²
	150 thick	195 kg/m²
	200 thick	260 kg/m²
Blockwork, insulating (560 kg/m³)	100 thick	60 kg/m²
	200 thick	120 kg/m²
Plaster or plasterboard, Gypsum.	13 thick	22 kg/m²
	16 thick	26 kg/m²

Bulk masses by volume

Water	1000 kg/m³
Sand, dry	1650 kg/m³
Earth, moist	2000 kg/m³
Concrete, stone (average figure)	2400 kg/m³
Softwood	500 kg/m³
Hardwood	700 kg/m³

APPENDIX II

SELECTED LIVE LOADS

from B.S.C.P. 3: chap. V; part I: 1967 amended 1970

	intensity of distributed load		concentrated load to be applied over any square of 300 mm side	
	kg/m^2	kN/m^2	kg	kN
Assembly buildings such as public halls, and theatres, gymnasia and hotel bars but excluding drill halls, places of worship, public lounges, schools, with fixed seating.	408	4.0	—	—
without fixed seating	510	5.0	367	3.6
Balconies	as the rooms to which they give access.			
Banking Halls	306	3.0	—	—
Bedrooms domestic	153	1.5	143	1.4
hotels and motels	204	2.0	184	1.4
institutional	153	1.5	184	1.8
Boiler rooms	765	7.5	459	4.5
Book and paper storage: figures given for each metre of storage height	245	2.4	714	7.0
Churches	306	3.0	275	2.7
Classrooms	306	3.0	275	2.7
Domestic dwellings	153	1.5	143	1.4
Garages: car parking for vehicles up to 2500 kg	255	2.5	918	9.0
workshops for vehicles up to 2500 kg	510	5.0	918	9.0
Hotel dining rooms and lounges	204	2.0	275	2.7
Offices: for general use	255	2.5	275	2.7
for data processing etc.	357	3.5	459	4.5
toilets	204	2.0	—	—
file rooms	510	5.0	459	4.5
Shops sales areas	408	4.0	367	3.6
Snow load, roofs	70	0.7	—	—

APPENDIX III

MOMENTS OF RESISTANCE: TIMBER
for timber joists of Grade 50 timber

based on $M = f \times \frac{bd^2}{6}$

size d × b mm	moment of resistance Nmm	moment of resistance kNm
50 × 50	104 167	0.10
75 × 50	234 375	0.23
75 × 75	351 562	0.35
100 × 37	308 333	0.31
100 × 50	416 667	0.41
100 × 75	625 000	0.62
125 × 37	481 770	0.48
125 × 50	651 042	0.65
125 × 75	976 562	0.97
150 × 50	937 500	0.93
150 × 75	1 406 250	1.40
150 × 100	1 875 000	1.87
175 × 50	1 276 042	1.27
200 × 50	1 666 667	1.66
200 × 100	3 333 333	3.33
225 × 50	2 109 375	2.11
225 × 75	3 164 062	3.16
225 × 100	4 218 750	4.22
250 × 50	2 604 167	2.60
250 × 75	3 906 250	3.90
275 × 50	3 151 042	3.15
275 × 75	4 726 562	4.72
300 × 50	3 750 000	3.75
300 × 75	5 625 000	5.62
300 × 100	7 500 000	7.50

APPENDIX IV

ROLLED STEEL JOISTS (203 × 102 and below)
Dimensions and properties

nominal size mm	mass per metre kg	web thick-ness t mm	area of section cm²	radius of gyration ry cm	moment of inertia cm⁴	modulus cm³	ratio D/T
203 × 102	25.33	5.8	32.3	2.25	2024	225.80	19.6
178 × 102	21.54	5.3	27.4	2.25	1339	170.90	19.7
152 × 89	17.09	4.9	21.8	1.99	762	115.60	18.4
127 × 114	26.79	7.4	34.1	2.63	834	148.80	11.2
127 × 76	13.36	4.5	17.0	1.72	400	74.94	16.7
102 × 64	9.65	4.1	12.3	1.43	182	42.84	15.4

APPENDIX V

ROLLED STEEL CHANNELS
Dimensions and properties

nominal size mm	mass per metre kg	web thick-ness t mm	area of section cm²	radius of gyration ry cm	moment of inertia cm⁴	modulus cm³	ratio D/T
305 × 102	46.18	10.2	58.83	2.91	6583	539.00	20.5
254 × 89	35.74	9.1	45.52	2.58	3612	350.20	18.7
203 × 89	29.78	8.1	37.94	2.64	1996	245.20	15.8
203 × 76	23.82	7.1	30.34	2.23	1508	192.00	18.2
178 × 89	26.81	7.6	34.15	2.66	1397	197.20	14.5
178 × 76	20.84	6.6	26.54	2.25	1028	150.40	17.3
152 × 89	23.84	7.1	30.36	2.66	924	153.00	13.2
152 × 76	17.88	6.4	22.77	2.24	653	111.80	16.9
127 × 64	14.90	6.4	18.98	1.88	367	75.99	13.8
102 × 51	10.42	6.1	13.28	1.48	167	40.89	13.3
76 × 38	6.70	5.1	8.53	1.12	54	19.46	11.2

Other sections than these are rolled but may be not all that available. Even some of the above particularly the larger sizes may have restriction of supply from time to time.

APPENDIX VI

UNIVERSAL BEAMS (203 × 133 and above)

Dimensions and properties

nominal size mm	mass per metre kg	web thick-ness t mm	area of section cm²	radius of gyration ry cm	moment of inertia cm⁴	modulus cm³	ratio $\frac{D}{T}$
457 × 191	67.0	8.5	85.4	4.12	26190	1296.0	35.7
457 × 152	52.0	7.6	66.5	3.11	19035	949.00	41.3
406 × 178	54.0	7.6	68.4	3.85	16457	925.3	37.0
406 × 140	39.0	6.3	49.4	2.89	11017	626.9	46.0
356 × 171	45.0	6.9	57.0	3.78	10609	686.9	36.2
356 × 127	33.0	5.9	41.8	2.59	7511	470.6	41.0
305 × 165	40.0	6.1	51.5	3.85	7384	561.2	29.9
305 × 127	37.0	7.2	47.5	2.67	6519	471.5	28.4
305 × 102	25.0	5.8	31.4	1.96	3962	287.9	44.6
254 × 146	31.0	6.1	40.0	3.35	3879	353.1	29.1
254 × 102	22.0	5.8	28.4	2.05	2575	225.7	37.2
203 × 133	25.0	5.8	32.3	3.10	2027	231.9	26.0

Within the above nominal sizes there is more than one size of Universal Beam rolled. The variation comes mainly in the thickness of the flanges which increase the the weight and strength of the beam. The above sizes are the lowest weight within each size and are the most popular and therefore most available. Steel tables from the British Constructional Steelwork Association or the major rolling mills provide full details of these other sizes. You may get offered these alternative weights per metre by suppliers or contractors. Within each nominal size the heavier rolling is stronger both in modulus and with increased radii of gyration and decreased D/T. If the above beams are satisfactory for you particular case then heavier ones will also be.

APPENDIX VII

REDUCED STEEL STRESSES IN BENDING

without adequate lateral restraint

TABLE 3a OF BS 449. ALLOWABLE STRESS

p_{bc} IN BENDING (N/mm^2) FOR BEAMS OF GRADE 43 STEEL

l/r_y	D/T							
	10	15	20	25	30	35	40	50
90	165	165	165	165	165	165	165	165
95	165	165	165	163	163	163	163	163
100	165	165	165	157	157	157	157	157
105	165	165	160	152	152	152	152	152
110	165	165	156	147	147	147	147	147
115	165	165	152	141	141	141	141	141
120	165	162	148	136	136	136	136	136
130	165	155	139	126	126	126	126	126
140	165	149	130	115	115	115	115	115
150	165	143	122	104	104	104	104	104
160	163	136	113	95	94	94	94	94
170	159	130	104	91	85	82	82	82
180	155	124	96	87	80	76	72	71
190	151	118	93	83	77	72	68	62
200	147	111	89	80	73	68	64	59
210	143	105	87	77	70	65	61	55
220	139	99	84	74	67	62	58	52
230	134	95	81	71	64	59	55	49
240	130	92	78	69	61	56	52	47
250	126	90	76	66	59	54	50	44
260	122	88	74	64	57	52	48	42
270	118	86	72	62	55	50	46	40
280	114	84	70	60	53	48	44	39
290	110	82	68	58	51	46	42	37
300	106	80	66	56	49	44	41	36

Intermediate values may be obtained by linear interpolation.

NOTE. For materials over 40 mm thick the stress shall not exceed 150 N/mm^2.

APPENDIX VIII

MOMENTS OF RESISTANCE: REINFORCED CONCRETE COMPRESSION

for concrete of 21·0 N/mm^2 cube crushing strength,

7·0 N/mm^2 design working stress

size d × b mm	moment of resistance Nmm	moment of resistance kNm
150 × 100	2 734 375	2.73
150 × 150	4 101 562	4.10
150 × 225	6 152 343	6.15
150 × 250	6 835 937	6.83
150 × 300	8 203 125	8.20
225 × 100	7 000 000	7.00
225 × 150	10 500 000	10.50
225 × 225	15 750 000	15.75
225 × 250	17 500 000	17.50
225 × 300	21 000 000	21.00
300 × 100	13 234 375	13.23
300 × 150	19 851 563	19.85

Note: The above figures have been calculated to allow for a lever arm 25 mm less than the overall depth shown as described on page 51.

APPENDIX IX

AREAS OF ROUND REINFORCING BARS

bar diam	area in mm^2 number of bars			
mm	1	2	3	4
6	28	56	84	113
10	78	157	235	314
12	113	226	339	452
15	176	353	530	706
18	254	509	763	1018

APPENDIX X

RECTANGULAR HOLLOW SECTIONS – selected

Dimensions and properties

nominal size mm	wall thickness t mm	mass per metre kg	area of section cm^2	radius gyration ry cm	modulus cm^3	ratio $\frac{D}{T}$
150 × 100	5.0	18.7	23.9	4.07	99.5	30.0
120 × 80	5.0	14.8	18.9	3.21	61.7	24.0
120 × 60	3.6	9.7	12.4	2.49	38.3	33.3
100 × 50	3.6	7.1	9.1	2.07	23.5	27.8
90 × 50	3.6	7.5	9.5	2.03	22.2	25.0
80 × 40	3.2	5.7	7.2	1.63	14.5	25.0
60 × 40	3.2	4.7	5.9	1.58	9.4	18.8
50 × 40	2.6	3.0	3.8	1.19	4.9	19.2

APPENDIX XI

SQUARE HOLLOW SECTIONS – selected

Dimensions and properties

nominal size mm	wall thickness t mm	mass per metre kg	area of section cm^2	radius of gyration ry cm
100 × 100	4.0	12.0	15.3	3.91
90 × 90	3.6	9.7	12.4	3.52
80 × 80	3.6	8.6	10.9	3.11
60 × 60	3.2	5.7	7.2	2.31
50 × 50	3.2	4.7	5.9	1.91

APPENDIX XII

CIRCULAR HOLLOW SECTIONS – selected

nominal size mm	wall thickness t mm	mass per metre kg	area of section cm^2	radius of gyration ry cm
76.1	3.2	5.7	7.3	2.58
48.3	3.2	3.6	4.5	1.60
33.7	2.6	2.0	2.5	1.10

APPENDIX XIII

REDUCED STEEL STRESSES IN AXIAL COMPRESSION

DESIGN OF STANCHIONS AND STRUTS

Axial stresses

The safe loads on stanchions and struts have been calculated in accordance with the allowable axial stresses specified in BS 449, Clause 30, and shown in the tables that follow:

TABLE 17a OF BS 449. ALLOWABLE STRESS p_c ON GROSS SECTION FOR AXIAL COMPRESSION FOR GRADE 43 STEEL

l/r	p_c (N/mm²) for grade 43 steel									
	0	1	2	3	4	5	6	7	8	9
0	155	155	154	154	153	153	153	152	152	151
10	151	151	150	150	149	149	148	148	148	147
20	147	146	146	146	145	145	144	144	144	143
30	143	142	142	142	141	141	141	140	140	139
40	139	138	138	137	137	136	136	136	135	134
50	133	133	132	131	130	130	129	128	127	126
60	126	125	124	123	122	121	120	119	118	117
70	115	114	113	112	111	110	108	107	106	105
80	104	102	101	100	99	97	96	95	94	92
90	91	90	89	87	86	85	84	83	81	80
100	79	78	77	76	75	74	73	72	71	70
110	69	68	67	66	65	64	63	62	61	61
120	60	59	58	57	56	56	55	54	53	53
130	52	51	51	50	49	49	48	48	47	46
140	46	45	45	44	43	43	42	42	41	41
150	40	40	39	39	38	38	38	37	37	36
160	36	35	35	35	34	34	33	33	33	32
170	32	32	31	31	31	30	30	30	29	29
180	29	28	28	28	28	27	27	27	26	26
190	26	26	25	25	25	25	24	24	24	24
200	24	23	23	23	23	22	22	22	22	22
210	21	21	21	21	21	20	20	20	20	20
220	20	19	19	19	19	19	19	18	18	18
230	18	18	18	18	17	17	17	17	17	17
240	17	16	16	16	16	16	16	16	16	15
250	15									
300	11									
350	8									

Intermediate values may be obtained by linear interpolation.

NOTE. For material over 40 mm thick, other than rolled I-beams or channels, and for universal columns of thicknesses exceeding 40 mm, the limiting stress is 140 N/mm².

APPENDIX XIV

BASIC COMPRESSIVE STRESSES – Brickwork B.S.C.P. III

TABLE 3*a*. BASIC COMPRESSIVE STRESSES FOR BRICKWORK MEMBERS (AT AND AFTER THE STATED TIMES)

Description of mortar	Mix (parts by volume)			Hardening time after completion of work†	Basic stress in MN/m² corresponding to units whose crushing strength (in MN/m²)‡ is:							
	Cement	Lime	Sand		2·8	7·0	10·5	20·5	27·5	34·5	52·0	69·0
				days								
Cement	1	0–¼*	3	7	0·28	0·70	1·05	1·65	2·05	2·50	3·50	4·55
	1	½	4½	14	0·28	0·70	0·95	1·45	1·70	2·05	2·80	3·60
Cement-lime	1	1	6	} 14	0·28	0·70	0·95	1·30	1·60	1·85	2·50	3·10
Cement with plasticizer	1	—	6									
Masonry cement	—	—	—									
Cement-lime	1	2	9	} 14	0·28	0·55	0·85	1·15	1·45	1·65	2·05	2·50
Cement with plasticizer	1	—	8									
Masonry cement	—	—	—									
Cement-lime	1	3	12	14	0·21	0·49	0·70	0·95	1·15	1·40	1·70	2·05
Hydraulic lime	—	1	2	14	0·21	0·49	0·70	0·95	1·15	1·40	1·70	2·05
Non-hydraulic	—	1	3	28	0·21	0·42	0·55	0·70	0·75	0·85	1·05	1·15

* The inclusion of lime in cement mortars is optional.

† These periods should be increased by the full amount of any time during which the air temperature remains below 4·4 °C plus half of any time during which the temperature is between 4·4 °C and 10 °C.

‡ Linear interpolation is permissible for units whose crushing strengths are intermediate between those given in the table.

APPENDIX XV

BRICKWORK STRESS REDUCTION FACTORS – B.S.C.P. III

TABLE 4. STRESS REDUCTION FACTORS FOR SLENDERNESS AND ECCENTRICITY

Slenderness ratio	Stress reduction factor*			
	Axially loaded	Eccentricity of vertical loading as a proportion of the thickness of the member		
		1/6	1/4	1/3†
6	1·00	1·00	1·00	1·00
8	0·95	0·93	0·92	0·91
10	0·89	0·85	0·83	0·81
12	0·84	0·78	0·75	0·72
14	0·78	0·70	0·66	0·62
16	0·73	0·63	0·58	0·53
18	0·67	0·55	0·49	0·43
20	0·62	0·48	0·41	0·34
22	0·56	0·40	0·32	0·24
24	0·51	0·33	0·24	—
26	0·45	0·25	—	—
27	0·43	0·22	—	—

* Linear interpolation between values is permitted.
† Where in special cases the eccentricity of loading lies between 1/3 and 1/2 of the thickness of the member, the stress reduction factor should vary linearly between unity and 0·20 for slenderness ratios of 6 and 20 respectively.

APPENDIX XVI

Mesh reinforcement

B.S. Ref.	pitch of of wires main	pitch of of wires cross	size of wire main	size of wire cross	cross sectional area per metre width main	cross sectional area per metre width cross	mass per square metre
A393	200	200	10	10	393	393	6.16
A252	200	200	8	8	252	252	3.95
A193	200	200	7	7	193	193	3.02
A142	200	200	6	6	142	142	2.22
A 98	200	200	5	5	98	98	1.54
B1131	100	200	12	8	1131	252	10.90
B 785	100	200	10	8	785	252	8.14
B 503	100	200	8	8	503	252	5.93
B 385	100	200	7	7	385	193	4.53
B 283	100	200	6	7	283	193	3.73
B 196	100	200	5	7	196	193	3.05
C 785	100	400	10	6	785	70	6.72
C 636	100	400	9	6	635	70	5.55
C 503	100	400	8	5	503	49	4.34
C 385	100	400	7	5	385	49	3.41
C 283	100	400	6	5	283	49	2.61
D 49	100	100	2.5	2.5	49	49	0.76
D 31	100	100	2	2	31	31	0.48